Environmental Innovation

Environmental Innovation

An Action Plan for Saving the Economy and the Planet by 2050

Jack Buffington

ROWMAN & LITTLEFIELD
Lanham • Boulder • New York • London

Published by Rowman & Littlefield
An imprint of The Rowman & Littlefield Publishing Group, Inc.
4501 Forbes Boulevard, Suite 200, Lanham, Maryland 20706
www.rowman.com

86-90 Paul Street, London EC2A 4NE

British Library Cataloguing in Publication Information Available

Library of Congress Cataloging-in-Publication Data Available

ISBN 978-1-5381-7814-0 (cloth: alk. paper)
ISBN 978-1-5381-7815-7 (ebook)

∞™ The paper used in this publication meets the minimum requirements of American National Standard for Information Sciences—Permanence of Paper for Printed Library Materials, ANSI/NISO Z39.48-1992.

Contents

Acknowledgments

I believe that humans are shaped by their communities, so I thank my parents, my wife Kari, and my daughters Kate and Marin for being the most important influences in my life. This book is dedicated to the process of asking questions rather than having the answers, because it is the former that will lead to solutions and the latter that will be an impediment to it.

Chapter 1

More Than Promises

Since its inception, Earth has been in a state of perpetual change. Biological changes to the earth have occurred due to solar cycles, natural climate fluctuations, volcanic activity, and even changes to its orbital wobble. Constant change has occurred, yet most of these natural events have been brief: limited short-term events that last only a decade or two. Over the last ten thousand years, the conditions on the planet have been remarkably stable, an understated fact that has been crucial for the social evolution of humanity. Yet today's anthropogenic changes to the earth have begun to alter this stability; to find a period with as much temperature volatility as today, you must go back to the last interglacial (the warm period between ice ages) age, around 125,000 years ago.[1] If not for temperature stability on the planet over the past ten thousand years, humans would not have evolved to such a great extent, but today, this stability is under a grave threat.

The stability of the earth's conditions enabled humans to exist homeostatically in the natural environment. For most of our history, we huddled within a narrow climate region characterized by mean temperatures of 30° Celsius (or 55° Fahrenheit).[2] The focus was on survival, as humans weren't the strongest or fastest, and we couldn't swim or fly. What humans learned to do better than others is collaborate in larger groups at high levels of sophistication. Advancements over human history through collaboration and cooperation have led to unimaginable levels of social and technological progress, working together within communities to create a built environment so significant that it is irreversibly changing the nature of the planet. Damage to the planet has also been extensive, potentially beyond our capacity (so far) to mitigate. Today, with carbon dioxide (CO_2) emissions at their highest levels in the last two to five million years, we face the Godlike challenge of reversing or slowing down our anthropogenic impact on the environment. In this brief period, CO_2 levels are 150 percent higher than they were before the Industrial Revolution.[3] By 2050, at the latest, we need to reverse engineer its aftereffects without damaging our societies and economies. We need to achieve these sustainability targets while at the same time advancing the economies of the developing world—no small task.

It's not politics or hyperbole to say the world faces an existential threat. The human race has never experienced a true global challenge and has no experience dealing with a matter as an international community. *Homo sapiens* have existed for approximately two hundred thousand years; human civilization, for six thousand. During these periods of development, our species has progressed not because we were the physically strongest or fastest but rather because of our collective intelligence and flexibility to solve problems as a group. But historically, "group" has been defined as local and tribal, not a global community. According to author Noah Harari, humans are the most flexible, adaptable species on the planet by far.[4] No other species can work collectively in large groups—groups where nearly everyone is a stranger to each other—like humans can. Chimpanzees can cooperate flexibly with a few others who are related, and ants can work across large groups, but in a rigid, programmed manner. Humans have the most remarkable social group intelligence capabilities by far. In his 2016 book *Tribe*, Sebastian Junger notes that "if the human race is under threat in some way that we don't yet understand, it will probably be at a community level that we either solve the problem or fail to."[5] Humans have primarily evolved through cooperation within tribes, with broader associations being less effective than we often understand.

Human collaboration and advancement have been done through the networking of tribes more so than by expanding a so-called global community. For example, extraplanetary travel to launch a rocket to the moon or Mars was national rather than a collaborative international strategy.

Intergovernmental bodies such as the United Nations (UN) have been limited, at best, in achieving global strategy for large-scale, transformational issues; one need only look back as recently as 2020 to see how the so-called global response to COVID-19 wasn't the case. There is a false belief—or maybe just a hope—that collaboration can happen within a global community to address existential problems such as climate change. Humans aspire to address these environmental challenges on a global scale through a global response, but no evidence exists that this is possible. Instead, we must fix the global problem of climate change within our human limitations. A 21st-century approach to sustainability must take a paradigm shift, looking at the problem through a different lens—and we don't have much time to figure this out.

Another problem with our response to climate change has been a limited focus beyond policy, science, and technology; we must also factor in psychology, sociology, politics, economics, and the supply chain, among other factors. I will discuss how science and technology may be the most manageable aspect to address. Approximately twenty-five hundred years ago, the Greek philosopher Socrates developed a perspective known as Socratic knowledge: that true wisdom understands what we don't know.[6] If we apply Socratic knowledge to the current problem of climate change versus a self-aggrandizing, Godlike view of the world, we can honestly understand our limitations in addressing these challenges. Could it be possible that humans are overestimating our ability to solve a problem through this so-called global collaboration? In this book, I will make a case that our strengths have become our weaknesses; we must ask more questions than thinking we have all of the answers. Inventing the technologies is relatively simple (I will also discuss this in the book); the more significant challenge is our lack of knowledge and experience in solving a global challenge and addressing the root causes of what is happening rather than the symptoms, such as CO_2 emissions. We must aim to solve these problems within our capabilities. We need a new, 21st-century approach to these environmental crises.

Today, we must establish an approach to balance policy, science, economics, and supply chain to achieve environmental goals by 2050; solutions must extend beyond policy, science, and technology to include more effective human systems. When COVID-19 was thrust upon the world in 2020, there was cooperation within the science community in sharing the genome map of the virus. Still, other than that, countries reverted to a statist model of competing for resources, including hoarding, racing to develop the first vaccine, and implementing travel rules and restrictions. Just one calendar year after the genome sequencing, the first COVID-19 vaccine was deployed, with numerous others becoming available in 2021. In 2022, multiple vaccines were in distribution, with more than one hundred more undergoing clinical trials. At the beginning of 2022, one year after the first COVID vaccines were released, 70 percent of European Union residents had received two doses, while Africa had received enough to fully vaccinate only 10 percent of its population.[7] As late as 2023, China was experiencing a major COVID crisis while most of the Global North moved forward, no longer considering the virus a significant health crisis. There were examples of cooperation, but tribalism surrounded them, posing as unhelpful competition in addressing a global pandemic. Without a global strategy and distribution system for the vaccine, the world population—including the vaccinated—is more susceptible to mutating variants that could lead to significant health risks. From a technical standpoint, the vaccine strategy was effective; from a human systems standpoint, we are struggling with the effects of COVID years later because of our lack of ability to come together as a collective species.

Understanding and accepting human limitations is foundational to solving global challenges rather than the aspirations of the impossible. Humans evolved through being tribal, not global. The COVID-19 crisis and large-scale wars in Europe, Asia, and Africa demonstrate that these limitations and geopolitical divides have not changed throughout history. Opportunities should be considered to network tribes or communities as a 21st-century model of globalization. Despite instantaneous communication across the planet using smartphones and Internet devices, a technical networking method—effective human collaboration beyond technology—remains surprisingly unsophisticated. Even within a nation such as the United States, we aspire to come together as a people, yet toxic ideological divides often seem more comforting. Theoretically, it seems logical that climate change and humankind's impact on the planet should offer an opportunity to collaborate through science, markets, and supply chains. Yet, from a practical standpoint, there is no evidence that our human systems can achieve these objectives.

How can the concentration levels of CO_2 in the atmosphere be caught up in politics? There are those who question the environmentalists and scientists regarding the legitimacy of the science or who tacitly agree with the science but not the proposed solutions. Skepticism and debate are healthy principles within the scientific method, but taken to an extreme, they can lead to dangerous fringe groups that question the motives of legitimate science and environmental policy. These people have raised a range of concerns about climate funding, the profit motive from green investments, and an alleged plot to create global governance controlled by the ruling elite. In a 2019 Pew Research Center study, only 39 percent of Republicans believed the US government was doing too little to protect the climate and environment; in stark contrast, a whopping 90 percent of Democrats didn't think it was doing enough.[8] Depending on what side of the debate they fall on on, people consider fossil fuels or solar power to be good or bad, and their opinions have little to do with the science and supply chain applications themselves. Differing points of view are essential in fleshing out solutions, but at some point strategies must be developed, debated, and implemented through rational motives. In this book, I will propose solutions that embrace the healthy aspect of the tribal nature of humans as the foundation of a 21st-century sustainability strategy.

Most people agree that the climate crisis is a significant problem that must be addressed. Further, most agree that climate change and other environmental problems caused by humans must be fixed within a finite period; here, I will define solutions to implement by 2050. Most agree that there has been little to no progress over the past decades. Focusing a strategy on an agreement rather than division can lead to action. Actions can be undertaken to bring communities together and become a networked global solution. Past and present efforts to achieve international consensus has been detrimental and a waste of time. The last five decades or so have seen thirty World Climate Conferences, twenty-eight

Conferences of the Parties, and the creation of the UN Intergovernmental Panel on Climate Change (IPCC), along with an uncountable number of government and nongovernmental organizations (NGOs), private corporate social responsibility and environmental and social governance programs, and a declaration that climate change exists signed by thousands of concerned scientists. These causes have been led by some of the most prominent politicians in the United States, including presidential candidates Al Gore and John Kerry, and some of the most famous businesspeople, such as Bill Gates. Virtually every political and business leader worldwide is on record as a passionate supporter of a climate strategy, with few in the mainstream expressing opposition.

For ordinary citizens, there are plenty of opportunities to get involved, such as joining or supporting one of the myriad nonprofit organizations focused on a specific facet of the problem. There are even opportunities to get involved passively, such as spending countless hours binge-watching climate change movies and documentaries on Netflix or traveling to France to visit a climate change theme park and hug a tree that tells us, "Watch out for fire, droughts, storms, floods, and other disasters. Stop deforestation. The earth will soon be dead and torn."[9]

But what benefit have we seen from these conferences and other grand gestures in the so-called global community? The data show that these global events and narratives have led to little—or even negative—progress toward our 2050 goals. Attempts to solve these environmental issues through international bodies and faux consensus have failed. The root cause of the problem is our human limitations. By providing a new approach to solutions by 2050, we will begin to make progress that hasn't occurred in the past fifty years.

Time is running out. As has been said, insanity is doing the same thing repeatedly and expecting different results. The world is missing its targets from the Paris Agreement to keep temperature rises below 1.5° Celsius for the balance of the century, with a rise already of 1.2° Celsius in eight years. A new approach to sustainability that reflects the healthy tribal nature of human evolution, focusing on communities and individuals rather than seeking a global community, is a feasible approach. Science, technology, engineering, math, supply chain, and economics must support community-based systems that work. Through individuals and communities, the principles of the scientific method can be embraced and not weaponized as a political football between warring ideologies. Science should not be denied, but questioned; we shouldn't replace commercial markets, but reform them. Science, capitalism, and human systems must be reformed.

These central themes of science and markets will be addressed as a foundation within this book to enable new solutions across the planet at the community and individual levels. The practices of economics and science need to be reformed rather than replaced. Czech philosopher Václav Havel states, "Modern science kills God and takes his place on the vacant throne. Science is the sole legitimate arbiter of all relevant truth."[10] For sustainable markets by 2050, science must lead to advancements in technology and society through healthy skepticism, not faith or defiance. Humans look for patterns when evidence is lacking. Individuals and civilizations have sought patterns for knowledge over thousands of years to explain floods, earthquakes, and other natural phenomena that continue today. As a method of discovery, science has been a game changer; in its form as a human system, it can face the same limitations as others. Science has uncovered numerous mysteries of the past but is increasingly becoming too much of a faith-based system as discoveries become more challenging. This happened during the COVID-19 pandemic when scientific methods were short-circuited in a rush to answer questions, which led to a loss of credibility; simultaneously, breathtaking progress was made on an RNA vaccine. Religion, with its lack of empirical evidence, requires its believers to take a great deal on faith; science should be different, requiring skepticism since nothing can be proved, only disproved. No authority—a priest, a rabbi, or a scientist—should own the truth. Any sense of comfort in conventional wisdom being "settled fact," even in the name of science itself, is the enemy of progress. This uncomfortable situation must be realized and addressed. We must constantly challenge the status quo to solve problems of this magnitude. Science and business both must adopt a continuous improvement mindset to address environmental challenges.

As the drivers of global strategy, international think tanks must also consider their effectiveness in addressing planetary problems. Take, for instance, the World Economic Forum (WEF), one of the most exclusive think tanks of establishment thought, which holds a meeting every January in Davos, Switzerland. In the words of the WEF, "Global momentum to tackle the climate crisis has been building. Progress has been made on almost every front, from bold corporate emissions-reduction targets and investors shifting away from coal to a surge of support for net-zero targets and a rising movement of youth activists from Uganda to India, culminating in Greta Thunberg being recognized as *Time Magazine*'s 2019 'Person of the Year.'"[11] However, these meetings have borne little fruit relative to keeping temperatures from rising significantly, keeping half a billion people from experiencing record heat levels, or increasing poverty and violence. According to a 2022 Institute for Economics and Peace report, seventy nine countries (40 percent) have become less peaceful since 2008.[12] Incremental progress is not helpful without transformational change; moving forward by a small percentage is insufficient toward the goal. Incremental change and innovation can be an opportunity cost that prevents sufficient progress. For those living in the most severe climate and economic crises, mainly in the Global South, these international conferences have accomplished little to improve their lives. These grand global institutions are failing those in the greatest need.

No reboot of a 21st-century sustainability strategy can occur without a greater focus on the Global South, the absolute epicenter of the planet's environmental crisis. For those who live in the most extreme poverty, defined as earning less than $2.15 a day, the ecological problem cannot be a day-to-day focus, even though they are the most impacted. While the number of people living in extreme poverty has been falling, from 36 percent of the world's population in 1990 to 9.2 percent in 2020, this means more than 700 million are facing economic crisis alongside the global climate crisis. And if the climate crisis continues to follow current trends, the number in extreme poverty will increase significantly—by 130 million over the next ten years—making a terrible situation even worse.[13] Unfortunately, most policymakers speak of poverty only in terms of its extremity, without discussing it in more relative terms. If we change the definition of poverty to earning $5.50 or even $10.00 a day, the world's population swells to 46 percent or over 70 percent, respectively.[14] Despite the economic development gains seen in recent decades, these data show that up to 70 percent of the world's population, or more than five billion of the world's inhabitants, live in a day-to-day struggle, making less in a day than the de facto hourly poverty wage level for the United States.[15] No progress will be made to heal the environment without integrating natural events—such as floods, rising sea levels, and CO_2 emissions—into the hidden global definition of relative poverty. A lack of understanding of the true impact of poverty is the greatest failure in understanding today's climate change challenges.

If the goal is to address the world's environmental challenges, the economy must be considered in the same conversation, primarily related to the more than 70 percent of the world's population facing economic instability. Ignoring this association is an example of cognitive dissonance: World leaders claim that the climate crisis is a global matter, but if over 70 percent of the world's population lives in relative poverty, how can addressing these instabilities not be the main focus of a climate change strategy? Hypothetically, let's suppose the countries of the world came together to address the climate crisis, and that humankind's impact on the planet must be handled by 2050, as the UN suggests. In that case, there is no escaping the truth that fewer than the 30 percent and of the world's population living under relative economic stability must be responsible for those of the 70 percent who are not. Expecting those who struggle to survive on a day-to-day basis to partake in environmental policy exercises is unreasonable. In this book, I will make a case for how 10 to 30 percent of the world population could accept this responsibility under different terms, as compared to today's approach to sustainability, which isn't acceptable. It may seem too heavy of a burden for us, but is there another option?

According to a 2021 Pew Research Center poll, people worldwide are concerned about climate change and willing to take some actions to do something about it, but they have low confidence in the effectiveness of public and corporate policies taken to address these challenges.[16] A new approach must begin by dispelling the false narrative that dictates we must choose between the environment and the economy. A choice between the two leads in prioritizing the former has resulted in a lack of material progress for the latter. Emphasizing climate change policy through challenging capitalism is a feckless model that has failed for the past fifty years and will continue to fail moving forward because a large percentage of the population lives under economic duress.

I define this false dichotomy—choosing whether the economy or the existential threat of climate change is the most critical concern—as green (economy) versus green (environment). The current state model of capitalism is responsible for the ecological damage that has accelerated over the past century. Yet this same economic system has also been responsible for a fourfold decrease in extreme poverty over the past thirty years, a remarkable accomplishment. Capitalism may, as it is claimed, be killing the planet, but it has also been the only system to pull millions out of poverty—and there are hundreds of millions more yet to save. If the goal is to reach carbon neutrality by 2050, the worst strategy is forcing the general population to choose between present-day survival and future environmental sustainability. This philosophy has failed for decades, and it will lead to the opposite of the intended environmental objectives.

Mahatma Gandhi said it best when he noted that "capital as such is not evil; its wrong use is evil. Capital in some form or other will always be needed."[17] With under thirty years left to achieve carbon neutrality, policymakers must balance the needs of the environment and the economy. Promoting a green and green strategy—doing both rather than one versus the other—is the only hope we have for moving toward these 2050 goals, especially in light of modern crises related to not just the environment but also food, water, war, and shortages of medicine, to name a few. We can no longer afford to propose academic solutions that will not work. An example of this false narrative of "green or green" is Richard Heinberg's 2012 book *The End of Growth: Adapting to Our New Economic Reality*, which insists that economics have failed us and that we must plan for life after growth.[18] Ten years later, and with an increasing fervor toward this false narrative, Matthew Huber's message in his 2022 book *Climate Change as Class War: Building Socialism on a Warming Planet*, as one might expect from the title, promotes a joint strategy of a worldwide class war leading to socialism as the means of addressing these environmental challenges. While it is fair to predict the possibility of class warfare if the climate crisis persists—leading to more acute food, water, and security issues that exact havoc on the Global South—capitalism must be reformed, not replaced, to achieve a balance of green and green. Humans must work within their limitations and challenge single-minded policies to become inclusive of people's and the planet's needs.

It is also important to note that today's institutional forms of capitalism and science differ significantly from the earlier intentions of both disciplines. As a result, there is a deep-seated public distrust about the effectiveness of these institutions. Even in the United States, the wealthiest nation in the world, there are rising fears and stress among the public with a basis of economic and environmental despair. According to the American Psychological Association, Americans are more stressed now than at any time since these polls were first conducted in 2007[19]—not just climate change but also other stressors, such as the COVID-19 pandemic, economic and geopolitical insecurity, inflation, drug addiction, rising crime, healthcare, and women's health issues, have raised the anxiety levels in the general populace to dangerous levels. America has faced challenging eras in the past, such as the two World Wars and the Great Depression at the onset of the 20th century. Yet I remember asking my grandmother (who was born in 1900) about these times, and she told me that while the times were tough, the people had optimism because they believed in America's public and private institutions and their communities to pull them through the crises. Today, the population in the United States and around the world does not have that same faith in their government, public institutions, and the private multinational corporations that run their supply chains regarding matters of environmental damage and economic equality. I believe the people will rally to support a viable strategy that balances the economy and planet, but not if it is led by large public and private institutions that have failed to act in their best interests in the past. The main objective of this book is to drive a change from the current state approach to sustainability to a new system that I'm calling "21st-century sustainability." It is a movement away from the large public and private institutions and empowering a new policy, science, and supply chain model, focusing more on individuals and communities, to solve these problems.

Good versus Evil?

The science shows that anthropogenic climate change is real, but that doesn't mean the conventional solutions proposed over the past decades are equally legitimate. An individualized, networked, community approach represents a paradigm shift toward the ways human civilizations has solved major problems in the past. However, there is no time—or even tolerance—to completely tear down the large public and private institutions that have failed to solve these problems; instead, a complementary human systems approach must be established. If our present-day human institutions are the problem, solving problems from a community perspective will lead to fewer factions and divisions within societies, not more. Alignment of sustainability strategies can be gained by empowering communities and individuals.

Anger and frustration provide the momentum toward tearing down existing institutions, which seems gratifying, but raw emotion will only lead to more contention and won't solve these problems. An article by Ana Levy-Lyons titled "The Banality of Environmental Destruction" posits that today's business leaders are not evil but rather terribly and terrifyingly normal, leading to seemingly rational business transactions that lead to the violence done to the planet.[20] This article is an analogy based on Hannah Arendt's 1963 book *Eichmann in Jerusalem: A Report on the Banality of Evil*, based on Adolf Eichmann's deportation to Jerusalem and his matter-of-fact explanation of what he did during the Holocaust as simply "doing this job." Levy-Lyons associates business leaders with Eichmann's "evil": the failure to act from an individual, conscious standpoint, instead undertaking institutional groupthink. Comparing the acts of a corporate manager focused on meeting consumer desire and shareholder profit to the genocide of millions of Jews and others is an unhelpful, emotional tirade. I know from experience that corporate decisions are too often narrowly focused on a profit motive and are driving unprecedented climate problems, but hyperbole and raw emotion will only create further division. Developing a balance between the economy and the environment should not lead to a "good versus evil" conversation; any motive in doing so will lead to more harm than good.

Another unintended and unfortunate example of this dichotomy is the characterization of and public fascination with twenty-one-year-old Swedish environmental activist Greta Thunberg, who burst onto the scene at age fifteen to challenge humanity about what's happening in our environment. It's not Thunberg who is the issue; indeed, we need activists like her to challenge conventional wisdom if we are going to address these environmental challenges, and I will discuss this in detail in the book. The problem is that she has become an unintentional meme in the political debate spanning ideological divides. When comedian Bill Maher launched a vicious tirade against Generation Z, he noted that Greta Thunberg may be the consciousness of this generation, but Kylie Jenner—who has hundreds of millions more Instagram followers than Thunberg—represents them as she flashes her wealth and flies in a private jet.[21] Rather than listening to and acting on her concerns, environmentalists have showered Thunberg with prizes, awards, designations, medals, and honorary degrees. She has even had new species of beetle, spider, and two types of snails named after her! But Thunberg—who in a 2020 interview noted her dislike for the numerous politicians and other leaders who seek to take pictures with her to give the impression they care about the envronment[22]—isn't looking for fame. The presence of these toxic debates that turn even well-intended young activists into memes is proof of the dysfunction of the current strategy and the limitations of human systems.

A big part of the problem is a lack of understanding of the environmental challenges beyond the taglines for change. Consider the often-quoted goal of achieving carbon neutrality worldwide by the year 2050. Google the phrase "carbon neutrality and United Nations" and you will find from this web search that it is referred to as "the world's most urgent mission by 2050."[23] And yet despite its promotion as the most critical goal over the coming decades, the discussion is a mile wide and an inch deep from the mainstream media in defining the problem itself and a plan for how to fix it. So many environmentalists and policymakers speak about "carbon neutrality" as our most important goal without understanding the details about what it is and what must be done. Memes, taglines, and Twitter wars aren't moving us forward, particularly when the issues aren't understood. In the public discourse today, there is too much debate about the concept and insufficient focus on understanding the root causes of the problem.

In straightforward terms, carbon neutrality is an equilibrium in the amount of carbon emitted and absorbed to and from the atmosphere, resulting in a net zero balance. From 2020 to 2021, CO_2 emissions increased by 6 percent, to 36.3 billion tons—the highest total in history and an increase of more than two billion tons annually. Presently, the world's natural carbon sinks of soil, forest, and oceans remove between 9.5 and 11 billion tons a year, or 26 to 30 percent of the target needed to reach carbon neutrality.[24] As the world's use of fossil fuels has grown and the natural carbon sinks have been taken down for the built environment, the ability of the atmosphere to manage carbon has been decimated. Therefore, over the next two and a half decades, the world's population needs to both stop the growth of carbon emissions on the planet *and* chop away at this 70 percent gap between carbon emissions and absorption—and there is only a very short time to do it.

In this book, I will discuss strategies for eliminating carbon emissions and increasing the natural carbon sink, the proper equation to solve the problem. We can achieve carbon neutrality through basic, structured problem-solving by defining the current state and the targeted goal, then analyzing the gap between the two. We must develop solutions based on scientific and economic details and discovery, a multidisciplinary green and green approach that includes policymakers, scientists, supply chain professionals, and others. Education of the general public must focus on the process and this carbon neutrality equation, which includes the need for energy to eliminate poverty beyond efforts to save the planet. The approach requires us to work collectively through interlocking communities of interest and differing opinions rather than giving in to the divisiveness of good versus evil.

Media outlets must become allies working toward solutions rather than profiting from sensationalism and partisanship; they must become supporters of the critical "third rail" of politics that addresses these topics. A sensational tagline like "carbon neutrality by 2050" or "closed-loop system for plastic by 2050" may catch the eyes of viewers and lead them to advocate for the cause—or not. Either way, it does little to educate the public and galvanize support toward a solution. In this book, I will provide detailed solutions that address the science, technological, supply chain, economics, and cultural/societal elements needed to solve these challenges. They are possible if we consider them from this perspective. I will discuss the current, sobering reality of where we stand and present audacious, innovative, and achievable goals to meet these challenges in a manner that can succeed to protect our oceans, rainforests, atmosphere, and the health of the world's population. Education is most important to inform the general population of what must be done and to create the change that needs to happen at the community and individual levels. It's a paradigm shift in strategy. We must be educated about the scope of the problem and use that knowledge to build solutions within our natural limitations. We can accomplish an environmental plan through communities and individuals, not global bodies.

Despite their best efforts, large public and private institutions over the last thirty years have failed to achieve an effective balance between science and economics, or green and green. For example, in 2015, the UN developed a list of seventeen Sustainable Development Goals (SDGs) as a scorecard to understand how to address these most significant challenges facing our population and the planet.[25] Ultimately, these initiatives, which are logical in concept, lack specifics for how to achieve these objectives and get financial support for implementation. That same year, the *Economist* magazine classified the SDGs as "worse than useless," suggesting that implementing the 169 targets contained within the goals—at the cost of 4 percent of the world's GDP—is pure fantasy.[26]

Break the Institutions

In the fields in which I have worked—including consulting, manufacturing, supply chain, and academia/research—the goal of a project, operation, or research is to try to break it, to disprove the current state. In this context, goals aren't an end state but a perpetual model of continuous improvement. When properly applied, the scientific method leads to a growth in our knowledge and understanding of the world rather than promoting conventional wisdom or a "settled fact" that defines an end state. In business, the methods most often used for continuous improvement are Six Sigma, Lean, Deming's Quality Principles, and the Toyota Production System. As the theory goes, the goal in business is to *break stuff* (we sometimes use a different word than "stuff"). If it can't be broken, the process or solution can survive until we can eventually break it to find a better alternative. Everything is considered breakable, and there are always better options if you keep looking for them. In science, the same principles should apply. For 231 years, Isaac Newton's theory of gravitation was considered a universal law, an unchallengeable truth, until Einstein's theory of general relativity challenged these assumptions. Newton's theory works just fine in many scenarios—like an apple falling from a tree—but it is not a universal truth, as it does not hold up when calculating Mercury's orbit around the sun, among others. The moral of the story for science and business is to advance knowledge and progress; no matter how confident we are of its viability, we should never accept that any principle is definitively true, at the risk of eventually being wrong. The concept of continuous improvement is constantly moving forward. Breaking something is a good, not bad, practice. Challenging the brightest and most established thinkers and leaders worldwide is essential and shouldn't be discouraged. Only through this method can humans address climate change and other environmental challenges. We must change our mentality and approach when facing an existential crisis.

Today, there is less of a culture of continuous improvement than is necessary; rather than breaking theories and ideas to improve them for the sake of the environment and economy, our public and private institutions too often seek power through knowledge rather than supporting the Socratic method of always questioning and that nothing is a settled fact. A focus on the answers rather than the questions has led to little to no progress toward climate change goals (such as carbon neutrality by 2050) and other critical targets (including economic targets) such as the SDGs. These large public and private institutions, most notably national and intergovernmental organizations (such as the UN) and the multinational corporations that run our supply chains, are too often based on a culture of authority and power, leading the population to lose trust in their ability to work in our interests. In the United States, a 2021 survey found low levels of confidence in its most important institutions, such as Congress (12 percent), big business (18 percent), technology companies (29 percent), and public schools (30 percent).[27] Likewise, the UN conducted a study in 2021 that showed a decline in the public's trust in government across three regions (Africa, the Americas, and Europe), with an average of a little above 35 percent.[28] Respondents to a 2021 Pew Research Center study found a favorable impression of the work of the UN, with an overall median of 67 percent, but these are somewhat skewed results because the study included only respondents in developed nations across Europe, North America, and Asia.[29] Finally, a 2022 Gallup survey asked US respondents how much confidence they had in big business, a question it has asked since 1973. In 1973, the percentage of the respondents who replied "very little" was 20 percent; by 2022 that figure had doubled, to 40 percent.[30] From these surveys and other sources, such as election results (including participation), protests, media contributions, and the like, it is clear that there is a growing distrust in the ability of our large public and private institutions to address the challenges we face in the 21st century. As is the case of a false narrative of green versus green, there is a false narrative discourse of "big government" versus "big business," with each side taking a default position in choosing one over the other. In the United States, there is a growing shift further toward the margins due to increasing disenchantment with capitalism (moving toward the left and socialism) and government (moving toward the right and libertarianism).[31] These responses may be less about being in favor of socialism, less government, or anarchy than a sign of growing disenchantment with large institutions and a feeling of powerlessness to do anything about them.

There needs to be a paradigm shift away from relying solely on the large public and private institutions to achieve economic and environmental targets by 2050. There is no evidence that the UN or multinational corporations, in their present forms, can deliver these goals by 2050. As much as the world appreciates the role of the UN, as is evident in the survey data, this global intergovernmental organization has little authority over its 195 member nations and can do little other than write nonbinding agreements that sovereign nations can and often choose to ignore. (There is, however, a new role for organizations such as the UN that I will discuss in this book.) The same is true for the European Union: Although it does hold more significant sway over its member countries, those nations are home to less than 10 percent of the world's population, and the union itself is broken into twenty-seven sovereign nations. Large multinational corporations, which are directly responsible for carbon emissions and many other human activities impacting the environment and the global economy, they are not directly accountable to the public good; instead, according to the business model of economist Milton Friedman, their greatest responsibility is to make as much money for their shareholders as possible.[32] Large institutions will be vital to solving the climate crisis, but their role will be in support of individuals and communities, not as the drivers of the bus. Each of us needs to take charge, 21st-century style.

A central theme of this book is that these institutions' 20th-century sustainability strategy, which has offered us the false narrative of green versus green, has not succeeded into the 21st century. Another central theme is the relationship between poverty, economic growth, and the lack of progress toward sustainability targets. The principles of the scientific method, supply chains, and entrepreneurship in capitalism must be reformed to lead to the paradigm shift necessary to meet these 21st-century challenges. Using an approach of collaboration through continuous improvement via networked individuals and communities, we can break conventional wisdom. It is an expectation that while our large institutions have been pillars of society, leading to stability and growth for centuries, they are increasingly becoming roadblocks to innovation in this new age. These institutions have been divisive engines for both those who loathe private enterprise failures and gravitate toward a socialized system and those who distrust the government and seek actions beyond and outside it. These fractures exist within and outside of sovereign nations as each country cooperates—or does not cooperate—with others. Unfortunately, a lack of cooperation across governments, whether related to a global pandemic or the climate crisis, is a formula for a lack of progress.

Today, there's a growing need to use a more unified approach in addressing these challenges. For example, the grand global climate protocols—including the early 21st-century Kyoto Protocol—have been less than adequate beyond their ceremonial value. Adopted in 2005, the Kyoto Protocol was classified as "legally binding" to reduce emissions to 5 percent below 1990 levels, but it applied only to developed nations and not to developing countries such as China, where CO_2 levels continue to rise. While this agreement could be considered a step forward, the math didn't work out well: In 1990, total CO_2 emissions were 22.4 billion tons, about twice as high as the carbon sink; since then, emissions have increased by 62 percent. Even if each developed nation reduced its emissions by 5 percent, as required by the protocol (and they did not), an increase in emissions from developing countries would have surpassed any gains achieved, as is the case today. The United States signed this agreement in 1998 but never ratified it and later withdrew entirely. Ten years later, the Paris Agreement was enacted, requiring all nations to set pledges for emission reduction and carbon neutrality by the second half of the 21st century. The United States withdrew and then reentered this agreement, while other countries never approved it—and in the seven years that followed, emissions rose by 22.6 percent, from 29.6 billion tons in 2015 to 36.3 at the end of 2022. Environmentalists applauded the Biden administration for reentering this agreement without understanding that it is worth little more than the paper it is written on.

What This Book Is About

Life is a battle between your cells and bacteria, and if you are healthy, the cells win and you live. However, when things go wrong, the bacteria win, and an infection in your body can lead to your death. In the late 19th century, scientists began to develop antibiotics such as penicillin, which can kill bacteria without killing us. Antibiotics became a game changer for healthcare in the 20th century, preventing infection and enabling modern medical procedures, such as organ transplants, that were not possible before. However, bacteria and even fungi can become resistant to these treatments, and the more we overuse or even misuse the drugs, the greater resistance is built, threatening our health and food production. This story of life that began more than a hundred million years ago between mammals and bacteria took a shift in the last hundred years due to scientific innovation, but now the microorganisms are fighting back. Today, the natural world is fighting back against our global industrialization, and humans are not used to coming together as a global community to encounter such a battle. Human limitations present themselves with the assertion that we "need to save the planet"; it's not the planet at risk but rather our chosen way of life and the lives of other species. Technology can contribute only to a certain degree. Humans do not need to save the planet, but we do need to save ourselves and other life because we are in charge. As such, the question is whether we can evolve to overcome our two hundred thousand years of evolution, transforming ourselves beyond every other species of life to change our nature.

In chapter 2, I will address how our public and private institutions have failed at their mission related to climate change. Starting in 1988—which marks the first year of a congressional hearing to address the matter in the US Senate, the creation of the IPCC, and the beginning of globalizaton—and moving forward to today, I will highlight a chronology of green versus green, laying out the environmental policy and economic path over the past thirty-five years to understand how the world got to this predicament. Unfortunately, over these decades of focus, we have lost rather than gained ground relative to the historic problem. This chapter will present a critical understanding of the environmental movement and the path of global economic development collectively, rather than looking at each in a vacuum.

Chapter 3 will take a closer look at the impact of climate change and economic turmoil in the Global South. Today, most of the perspective related to the problem is from the developed world, while much of the devastation and many of the limitations have fallen on those living in places like Manila, Jakarta, or Dhaka. The developing world perspective is one of the greatest untold stories, not just for the sake of the billions who live in those parts of the world but also for the rest of us. In the 2020s, significant flashpoints in the Global South are ticking time bombs whose explosion will have severe consequences for the entire planet. Without undertaking best practices for both ourselves and those of the 70 percent who cannot focus on it, we have no chance of reaching our 2050 goals. Looking at the many public and private pledges in place today, we can see how these siloed efforts will not add up to what is needed in the final calculus if a large section of the planet is out of control.

Science is crucial to inventing the solutions required to address climate change, the effects of other human activities, and the misapplication of the scientific method perpetuated when present theories and proposed solutions are classified as settled fact. Transformational advances in chemistry, material science, biotechnology, life sciences, and others will be vital to meeting these 2050 goals. But can today's scientific practices help us achieve these goals? Chapter 4 will provide evidence for how to solve the problem. Today's model of big science undertaken by big institutions has been deficient in challenging the status quo and addressing the collective existential threats related to the environment. Science as a formalized process using the empirical method can achieve these objectives within the time frame required, but not if the scientific community continues, as Václav Havel noted, to consider itself the sole judge of all relevant truth. This chapter will closely examine the dogma of current science and the institutional applications of big science to determine how it can be reformed—and quickly.

Chapter 5 will focus on the global supply chain, which often conflicts with the public good. For instance, to produce enough high-quality clothing to meet the needs of the world's population at a profit, multinational corporations must enable long-tailed supply chains that are harmful to the environment. Similarly, 50 percent of the world's plastic is single use, which both increases profitability and causes environmental waste and damage. Global supply chain systems must be transformed to meet our 2050 global objectives. Since the beginning of the COVID-19 pandemic, the world has awakened to the limitations of our present-day supply chain, but the roots of the problem, which existed even before COVID-19, must be addressed.

In chapter 6, I will begin to focus on what needs to happen to develop solutions for carbon neutrality by 2050. In this chapter, I will lay out the math of the challenges we face in reaching this result in under thirty years. Today, 85 percent of the world's energy is based on fossil fuels: oil, coal, and gas. This energy formula and global economic growth are why CO_2 emissions are three times higher than the natural carbon sink and grow by 6 percent annually. Furthermore, the world is losing ground in solving this equation as a result of major global flashpoints such as the COVID-19 pandemic and the wars in Ukraine and the Middle East. As a result, the challenges we face in reaching this goal are more complex but still achievable. Using math as a basis, I will present a feasible plan for reforming our energy markets to reach carbon neutrality by 2050.

The food production system accounts for approximately 18 percent of greenhouse gas emissions; 60 percent is due to meat production. Chapter 7 will address the challenges associated with the food systems needed to feed a growing world population approaching eight billion. In the late 18th century, Thomas Malthus raised the issue of a population growing much faster than the food system, but technological advancements proved him wrong—at least for the time being.[33] As a more significant percentage of the world's population emerges from extreme poverty, there's a growth in the demand for meat, which concerning as it is one of the most unsustainable activities on the planet. In addition, increases in meat production and consumption are leading to a growth in other challenges, such as the growth in infectious disease spread and the misuse and overuse of antibiotics in livestock production that impact their effectiveness moving forward. Water supply may present an even more significant challenge in the future, given the nature of historic droughts worldwide. The good news is that scientific and supply chain innovations, if appropriately and rapidly implemented, can effectively address these challenges. The question is, can these solutions happen quickly enough? I will discuss the game-changing solutions that need to be implemented as swiftly and effectively as possible to beat the clock on climate change and its effect of human activities.

Life in the 21st century requires a lot of natural and synthetic materials. Some are finite raw materials—renewable and nonrenewable—from our planet, while others are synthetic and lack transparency in the production and supply chain process. Perhaps even more so than energy and food is the issue of materials in our lives: invisible to our focus but everywhere we look. These materials include, but are not restricted to, cement, steel, plastic, fertilizer, wood, and a near-infinite combination of subcategories, additives, and so on. Of note is the use of chemicals that are difficult, if not impossible, to regulate across the global supply chain; it is therefore nearly impossible to verify their safety before their use. Another focus topic is plastic, which the UN has estimated will be more plentiful than fish in the ocean by 2050. In chapter 8, I will discuss how materials must be designed, manufactured, distributed, and sustained differently to meet our 2050 goals.

The most significant challenges we face in the future are related not to energy, food, water, or materials but rather to our human limitations, as I have discussed briefly in this chapter. All life forms seek to expand to their natural limitations, and in every case where this happens in an ecosystem, it has led to peril as a result of the lack of resources to sustain growth. To some extent, humans have been an exception to this rule because we have been the only species thus far able to control the resources on the planet. But as we are now learning, this has its limits: The natural environment is pushing back. Still, the question is, are humans unique in that we can we evolve sufficiently to take on these challenges? Can we save ourselves from ourselves? Or can we, perhaps, accept and achieve moderate growth through cooperation across cultures rather than competition through science and the supply chain? The answer isn't clear, but a new design using technology to enable communities and individuals, rather than large institutions, as the foundational solution may provide us with a better opportunity than a model run through the large public and private institutions that have led to the problems we are facing today. I will discuss this critical topic in chapter 9.

The concluding chapter of this book will discuss how we can put these concepts into action within a time line, a plan by 2050, to better balance the economy and the planet. First, I will provide details in sequence for how we must meet these challenges, then I will give a detailed and practical roadmap for making them happen and achieving our climate goals and other human activity objectives by 2050. This blueprint balances the limitations and capabilities of science, supply chains, economics, and human systems.

The purpose of this book is to serve as a catalyst for change. Most of us in the developed world understand and accept the problems related to the environment, even if we do not understand the full implications of how to balance these concerns, such as economic insecurity, human rights, and other social situations, with others. The challenges we face in the 21st century can be overwhelming: for those in the Global South, as an immediate crisis of survival, and for many in the Global North, as a more conceptual and existential humanitarian crisis. The wealthy nations need to develop effective green and green solutions for ourselves and, even more importantly, for those who cannot do so for themselves.

I think there's hope—but it involves a paradigm shift toward greater individual and community control over the problems and the solutions, rather than a feeling of individual helplessness and the expectation that large institutions such as the United Nations, national governments, and multinational corporations will solve these problems. Large institutions are not intentionally causing these problems, but they cannot optimize what's best for the individual and the environment because their sphere of influence is too broad and nonaligned. Perhaps a new capitalist model can create the right balance where technology optimizes and enables these decisions locally. This book provides evidence for a new approach to replace what hasn't been effective over the past thirty years. So let's get on with it!

Chapter 2

21st-Century Environmental Policy

Henry David Thoreau was a 19th-century writer, naturalist, and philosopher best known for his classic 1854 work *Walden; or, Life in the Woods*. Thoreau also published an essay titled "On the Duty of Civil Disobedience," written a day after he was jailed for not paying his poll tax, an obligation due every six years. Thoreau popularized the concept of civil disobedience, which was used a century later by leaders such as Martin Luther King Jr. and Mahatma Gandhi. Thoreau is revered by environmentalists, who take pilgrimages to Walden Pond to swim in what he described as a crystal-clear bottomless vestige of nature. Yet the folklore is different from modern reality. If you visit Concord, Massachusetts, today, you'll find Walden Pond feels like a typical overcrowded museum, including a dangerously high concentration of human urine in the pond as a result of so many pilgrims relieving themselves while swimming.[1]

Ted Kaczynski—popularly known as the Unabomber—fancied himself a modern-day Thoreau. Inspired by Thoreau's writings, Kaczynski moved to a remote cabin in Montana to live self-sufficiently away from civilization in a form of green anarchism. Both men retreated from society to remote cabins to practice self-reliance and inner peace, but Kaczynski also sent homemade bombs through the mail, killing three people and injuring twenty-three others. Kaczynski's essay "Industrial Society and Its Future," also known as the "Unabomber Manifesto," was published in 1995 by the *Washington Post* in a deal to halt the bombings.

Thoreau's novel *Walden* teaches us to live deliberately within the essentials of life, while modern-day environmental activists can sometimes be frustrated to the point of destruction and death. According to modern offshoots of the environmental movement, such as the Earth Liberation Front, economic sabotage and warfare are justified to save the planet from humankind's destruction. In 2021, Swedish professor Andreas Malm wrote *How to Blow Up a Pipeline: Learning to Fight a World on Fire*, which justifies the destruction of property to protect the people. The lack of substantive progress driven by the large institutions in charge has unintentionally led to increasingly dangerous acts of frustration that can never be justified but are increasingly rationalized.

Few topics are more important to environmentalists than their connection to nature. Environmentalists and biologists believe the relationship between nature and humanity remains vital and critical to our evolution. Renowned biologist Edward O. Wilson—often referred to as the "ant-man" because of his lifetime study of the insect—developed the "biophilia hypothesis" asserting the innate human connection to nature given our evolutionary history over more than two hundred thousand years as a species.[2] The human connection to nature increases success in finding resources, security, and other keys to survival. We live in a built society, but we aspire to be a part of nature, even though we no longer need to do so and it is no longer possible. Therefore, Wilson's biophilia hypothesis is rendered obsolete.

In 1989, environmentalist Bill McKibben published *The End of Nature*, which announced the end of nature's reign over the planet and described how industrial activity and human disconnection from the earth are destroying the environment. McKibben's book is considered the first to identify the problem of climate change for a general audience, explaining that nature's increasingly severe natural events are driven by anthropogenic activity. To McKibben and the burgeoning environmental movement, this "end of nature" is a doomsday event leading to the ultimate destruction of the planet. To others, the built environment is the system that allows humans to thrive. The contention that exists is the disconnect between these two worlds.

McKibben emphasized this disconnect within two responses to this "end of nature." The first is a "defiant reflex," a reluctance to change given the conflict between our modern comforts and their impact on the planet.[3] The "choice of doing nothing" and burning more fossil fuels has been the course of action for more than thirty years, resulting in a 65 percent increase in global fuel consumption, from 82.2 terawatt hours in 1989 to 136 terawatt hours in 2021.[4] McKibben proposed a second option: a "humbler way of life" where humankind is no longer in control of nature.

Suggesting a binary choice between no change in our modern lives or a "humbler" way of life does not understand the bigger picture across the planet, particularly given that so many live in the humblest way possible. When McKibben published his book in 1989, the world population was 5.3 billion, with 1.9 billion—35.8 percent—in extreme poverty. In 2022, the world population just surpassed 8 billion, with approximately 667 million, or 8.3 percent, in extreme poverty.[5] A significant number of people in the world no longer live in the worst form of poverty, but hundreds of millions still do. For environmental policy to be effective, it must be nuanced to the present-day reality for most in the world (even after this progress) that daily life is a constant struggle.

Our Fractured Planet

According to environmentalists, we should live as one family within nature, but the truth is that we do not live this place; we live on a fractured planet of localized flashpoints and structured communities, for better and worse. Even concerning global climate patterns, humans and every other species face survival locally, not globally. For example, over the past decades, part of the planet has become greener through the CO_2-enriched atmosphere, and northern climates have become crop-growing regions. In contrast, areas once conducive to plant life have become browner, and the higher absorption of CO_2 into oceans and other bodies of water has been harmful. Throughout history, survival has been threatened local by wildlife, infectious diseases, natural weather events, starvation, and other dangerous circumstances.

In many cases, these local disturbances led to human, animal, and plant migration patterns, which led to one region impacting another. In the 1970s, chemist James Lovelock and microbiologist Lynn Margulis developed the Gaia hypothesis, positing that organisms and their nonliving surroundings interact synergistically to maintain earth conditions. Today, a Gaia hypothesis exists in a different pattern: organic and inorganic elements are impacting one another in a chaotic and negative synergistic manner through climate crisis migration, infectious disease pandemics worsened by globalization, war, and other scourges. Gaia may be the mother of life, but human societies have taken over in a nonsynergistic manner regulated through a near unlimited number of societies and municipalities in competition, not cooperation. There is no universal organizing paradigm to keep the planet in balance.

Environmental policies have failed when they focused on this unrealistic concept of Gaia, the mother of life. Lovelock's principle is indeed beautiful and inspiring, but it focuses on the symptoms, rather than the root cause problems for issues, such as the impact of high concentration levels of CO_2 in the atmosphere and plastic waste in the oceans. Concepts such as the Gaia hypothesis fool us into believing our social structures and dominion over the planet are akin to how nature worked before we took over. The promise of an integrated and complex system of Gaia through human systems such as the UN cannot address the reality spanning 195 designated nations worldwide and perhaps as many as thirty-five hundred to five thousand distinct municipalities defined as states, counties, towns, cities, and so on. Each nation is sovereign, and so are many states and local municipalities. Sovereign entities exist as countries, cities, and towns, but they also exist as undesignated tribes with unique differences in culture, history, socioeconomics, and ideology. Gaia isn't displayed when a majority of the benefits from industrialization are reaped by those in the Global North while those in the Global South are paying the highest environmental price. According to a 2022 S&P Global risk report, Asia and Africa face much more significant risks due to climate change than Europe and the Americas do. Moreover, the risk disproportionately impacts lower-income rather than higher-income areas.[6] Our tribal nature and success through social evolution is antithetical to Gaia; environmental policy may wish for a global community to emerge through this spiritual relationship to the planet, but how much longer will it take those who lead these efforts to realize the futility of such measures?

Human geography demonstrates the lack of global governance over the planet. The ocean covers 70 percent of the earth; 43 percent of that—and 95 percent of the ocean's volume—is international waters and the remaining 27 percent territorial waters.[7] These numbers add up to a planet where 43 percent of the surface area is not the responsibility of any single nation. Taking the 27 percent of the territorial waters and the 2.75 percent of landmass that is Antarctica, not the property of any one nation, out of the equation leaves approximately 27 percent of the entire planet up for grabs among the 195 nations in the world. Of this available land, the largest landmass nations are Russia, at 3.34 percent of the total (or 12.3 percent), and Canada, at 1.96 percent (7.2 percent). The math is clear: a global climate strategy is impossible when so much of the planet isn't under jurisdiction, and the small amount that remains is fought over for resources across 195 nations and their large number of sovereign municipalities.

Not only is there no unified governance over the planet, but much of the planet remains not understood, despite technological advances. An estimated 80 percent of the ocean—comprising more than half the planet and much more significant in volume—has not been mapped or explored. No nation is responsible for the atmosphere or even understands what it truly means to contain the rise in temperature to 1.5° Celsius, as agreed within the Paris Agreement. Nobody is responsible for the Antarctic, and while parts of rivers are sovereign property within national boundaries, many extend between nations, leading to both cooperation and competition. The Amazon rainforest spans eight countries, while eight nations claim sovereignty in the Arctic and seven in the Andes Mountains.

Closer to home in the United States, the Colorado River spans more than fourteen hundred miles across five US and two Mexican states, providing water to 40 million people. Despite the importance of cooperation to ensure the river's health, it faces overconsumption and a drought that is historic in over a thousand years. Each of these jurisdictions has water rights laws, some focused on seniority, and a "use or it lose it" model that essentially incentivizes users to waste water to maintain their rights. These laws are focused on the rights of the individual, not as a representative of a Gaian approach on behalf of the planet. Even within a sovereign nation, establishing policy has become difficult when the natural system is integrated and connected while the human system is fragmented and flawed.

In contrast, the Indigenous peoples who inhabited these lands—the Pueblo and Navajo—considered the river sacred, owned by the earth rather than humans. Today, environmental rights attorneys from every municipality in the world fight over who owns what and bears responsibility for nothing. The world of policymakers and attorneys map rules for sovereign and private self-interest, multinational corporations who hold no sovereignty, and numerous other justifications without deference to the concept of a Gaia hypothesis. On a planet with eight billion inhabitants, we no longer live within the land in the manner that McKibben advocates. And yet, environmental policy either believes or wishes to hope that such a concept can work to save the planet. A 21st-century approach to environmental policy must factor in these realities, or we will continue to head toward a path of our destruction.

Saving the Planet?

Starting in the 1960s with the hippie environmentalist movement, Indigenous North Americans were considered "America's original environmentalists," a simplistic trope based on historical stereotypes. The concept of the "environmental Indian" stems, at least in part, from seeing few differences between the natives and wild animals, a view once used to rationalize exterminating a population considered to be subhuman. The view that Indigenous peoples were one with the land led to a lack of acknowledgment of their sophisticated land management practices and other innovations left behind after settlement.[8] For example, 19th-century environmentalist John Muir said, "Indians walk softly and hurt the landscape hardly more than birds or squirrels."[9] In 1892, Muir founded the Sierra Club, which would grow into one of the largest environmental advocacy groups worldwide, to "make the mountains glad" and start an environmentalist movement aimed at taking a "soft touch" to the natural world rather than the solutions to managing it practiced by Indigenous peoples. Another well-known naturalist and conservationist, Aldo Leopold is known for his 1949 book *A Sand County Almanac*. In this book, Leopold defines a "land ethic" as a moral code of conduct for how people should care for the land, which will care for us in return. These first-generation environmentalists, including Thoreau, practiced conservation by living on the land with reverence and affinity, in some ways like the Indigenous peoples. As the built environment drove us away from nature and toward cities and suburbs, our notions of environmental policy became more conceptual than experiential and engineering based. In concept, an environmentalist holds Indigenous peoples in high regard without appreciating the experiences and sophisticated engineering techniques they employed to manage this homeostatic balance. The tropes developed by environmentalists understand neither the experiences and technical skills of the Indigenous peoples nor the realities of today, given economics and demographics. Environmental policy based on emotion and conceptual ethics is often well intended from afar but fails to understand the practicalities on the ground.

In the 1960s and 1970s, with the start of the counterculture movement in America, the strategy of environmentalists was based on legal and moral codes tied to their strong beliefs in humanity's inner connection to the land. In this model of nature worship, it was conceived that humans should leave nature alone and stop acting immorally by wreaking havoc on it. These views are in contrast to those of Indigenous peoples, who are sometimes stereotyped as primitive savages but practiced sophisticated engineering solutions in connection to the land through living sustainably within it. While the Indigenous peoples lived in tribal cultures that respected the land and viewed rivers as owned by the planet rather than people, the environmentalists worked in the counterculture movements to fight those in power and to treat nature as some form of a living museum. In 1970, the first Earth Day was a protest of approximately 20 million Americans; today, it boasts up to a billion global participants, according to the Earth Day website.[10] Earth Day was the beginning of the movement of environmentalism toward morality, legalities, ethics, and emotion and away from engineering and structured problem-solving; the former is well intended but ineffective, while the latter is practical and effective, and could be so in the future.

Noted biologist Rachel Carson, who wrote the 1962 book *Silent Spring*, sounded the alarm about the application of dangerous synthetic chemicals and pollution in the United States. Carson was a scientist who provided significant evidence that environmental policy needed to address these problems from a legal and regulatory standpoint based on science and engineering. Environmental laws and regulations were nonpartisan issues, and the Environmental Protection Agency (EPA) was created during the Republican administration under Richard Nixon. Throughout the 1970s, Congress passed multiple environmental bills, including the Clean Water Act of 1972, the Safe Drinking Water Act of 1974, the Toxic Substances Act of 1976, and the Resource Conservation and Recovery Act of 1976. In the winter of 1977, President Jimmy Carter sought to rally the American public by giving a speech in the White House wearing a cardigan sweater and with the heat turned down. In 1979, Carter installed thirty-two solar panels on the roof of the White House. However, while this was happening, the United States was facing one of the worst economic crises in its history, and the populace that was facing massive inflation and a falling standard of living perceived environmentalism as the enemy of economic growth. Ronald Reagan took over as president in 1981 with a focus on business, and starting that year the EPA and environmental policy became a political football game between the Democrats and the Republicans that continues today. One side—the Republicans—viewed the EPA as an entity that inhibited growth through bureaucratic practices, and they were correct. On the other side, the Democrats viewed the EPA as one of the few organizations that could offset the power of the large corporations that have damaged the environment with impunity, and they were also correct. And so it began that environmental policy became an ideological talking point rather than a practice to focus on both the economy *and* the environment.

After Congress passed regulations in the 1970s and 1980s with bipartisan support, partisan politics once again prevailed. In expression of his opposition to the Clean Air Act in 1995, Republican congressman Tom DeLay said he did not favor a "Gestapo-type government imposing regulations on the American public."[11] On the other side, liberal economist Paul Krugman said that opposition from the right to climate change measures isn't about economic growth as much as it is a culture war driven by race and ethnicity.[12] This ideological divide related to environmental policy may have been birthed in the United States, but it exists worldwide today. As a result, environmental policy has moved to the political arena, far away from engineers, scientists, and supply chain practitioners. Divisive politics have overtaken communities as a method of solving problems.

Red Light, Green Light

Nobody—not even his supporters—considers Donald Trump a champion for the environment. To some, this is a good thing; to others, it is a tragedy. The Trump administration rolled back more than one hundred regulations through the EPA and the Department of the Interior.[13] Many of these rollbacks have led to many legal battles. Some have been reinstated, leading to further turmoil surrounding environmental policy led by attorneys, not scientists and supply chain practitioners. While some rollbacks challenge onerous government regulations that usurp economic growth, a viable position, others seem politically motivated or even downright ideologically based, serving little purpose other than to overturn whatever the prior administration put into place. Eliminating barriers to innovation is good, but rolling back laws that protect the environment for political benefit isn't; the rational means to protect the environment and the economy in a bipartisan manner, as occurred in the 1970s, has been replaced with ideological turf wars.

One aspect of Trump's environmental strategy was simple: to clog the public policy discussion and courts with debates to overturn past regulations to prevent further, even more restrictive policies from moving forward. If environmental public policy experts are busy defending their existing turf, after all, how can they promote new legislation? Politicians have turned the environmental policy into a game for control, driven by policy analysts and attorneys rather than scientists and supply chain practitioners. Because some Americans must focus much of their energy on day-to-day life matters, they do not have the time or patience to understand environmental policy. If you ask the average American about Trump's environmental record, they will likely mention not the policies that influence rivers, public spaces, or a focus on coal production in the United States but rather those related to splashy global agreements such as the Paris Agreement. Because every nation signed it, the Paris Agreement is seen as landmark legislation, including a coming together of President Barack Obama of the United States and Xi Jinping of China. Rather than focusing on environmental rollbacks that are proved as harmful to the environment, environmentalists focused Trump's record on the one agreement that had literally no impact other than symbolism for the movement. There is no evidence that the Paris Agreement is making a difference on people's lives or curing the impending environmental crisis.

In the details of Trump's environmental policy record are stories of unabashed and petty partisan politics that lack a cohesive understanding of science and economics. A majority of environmentalists have spent little time questioning some of these details and have been more focused on the pulling of the Paris Agreement as the Trump administration's greatest sin. And yet, despite the pomp and circumstance of this signatory global agreement, as it stands today, at the end of 2022, the United Nations has itself calculated that there is a 93 percent probability that the 1.5⊠ Celsius cap on rising temperatures will be surpassed within the next five years, much earlier than the 2100 goal.[14] Therefore, Trump's challenge to the effectiveness of the Paris Agreement was correct, even if done for the wrong reasons.

In June 2017, Trump announced the US withdrawal from the Paris Agreement, however given the bureaucracy in the agreement made it difficult to withdraw, that didn't happen until November 2020, at the end of Trump's presidency. Nearly three months later, in January 2021, the Biden administration announced that the United States would rejoin the agreement. In October of that year, five years after it was announced, Turkey became the last nation to ratify the agreement. Today, according to the climate watchdog group Climate Action Tracker, none of the major economies, including the G20 nations, are meeting their obligations per the Paris Agreement.[15]

The Paris Agreement is just one example of an approach to environmental policy that focuses on law, ethics, and emotion rather than engineering and structured problem-solving. The first-ever global environmental assessment conducted by the United Nations in 2019 found a thirty-eight-fold increase in statutes enacted worldwide since 1972 but failure to implement or enforce these policies.[16] According to Yale environmental law expert Daniel Esty, the problem with environmental law as it has been practiced since its inception in the 1970s and 1980s is its focus on the government telling businesses what they can't do, which leads to a back-and-forth focus on the environment versus the economy, as is shown in the example above.[17] This "red light" approach, developed in the early days of the Industrial Revolution when kerosene producers dumped flammable toxic waste into rivers, made sense back then. This approach is needed in developing parts of the world but can become prohibitive and bureaucratic in the face of economic and environmental improvements. Since the 1980s, the prevailing notion has been that environmental policy and trade should be separate—as if this is a viable strategy. In Esty's words, we need an environmental policy approach focusing on fewer (not zero) red lights and more green lights. It might be logical for a body such as the World Trade Organization to work toward balance between the two rather than maintain the current network, in which some organizations are responsible for pushing for and managing the red lights while others advocate solely for the green lights.

Institutional Failure

It's time to acknowledge the failure of environmental policy that focuses on large institutions to address these balance challenges. When discussing "large institutions," I am speaking about intergovernmental organizations such as the UN; national governmental agencies such as the EPA; large NGOs such as the Nature Conservancy, Greenpeace, and Earthwatch; and efforts within large multinational corporations, including corporate social responsibility (CSR) and environmental, social, and governance (ESG) programs.

The first organization to discuss is the UN Environment Programme (UNEP). Founded in 1972, the UNEP is focused on awareness and advocacy to address environmental action. One of its endeavors is the IPCC, viewed worldwide as the global authority on climate-related matters. It does not conduct research, however, but is an independent reviewer of scientific research conducted worldwide; by not conducting its own research, it is too slow to respond, although it is considered the foremost authority on climate change. A problem exists when an institution like the IPCC, which lacks both credentials and authority, is regarded as a worldwide expert on climate matters. The ineffectiveness of the IPCC is an example of the incapability of large institutions to lead change for a global challenge.

Public agencies such as the UNEP and the EPA are often classified as ineffective and underfunded without recognizing that funding impacts effectiveness. In 2019, at the behest of scientists worldwide, *Nature* magazine published an article calling for reform of the UNEP, noting its patchwork of environmental agencies and institutions, which are often accused of special interests and cronyism and do not possess enough authority and funding to do the job correctly. According to this article, UNEP funding decreased by 37 percent between 1979 and 2019, except for some special projects that have seen substantially increased funding.[18] This was not the first time reform or even reorganization was proposed for the UNEP. In 2007, French president Jacques Chirac's call for reform of the UNEP, which would have created a more powerful body called the United Nations Environment Organization, was endorsed by forty-six nations but blocked by the United States, China, Russia, and Saudi Arabia. Kenya, where the current UNEP organization is headquartered, accused the French of an end-run seeking to wrest control of the organization.

Similarly, the EPA was created in 1970 based on bipartisan and public outrage related to the pollution of American rivers and skies. Back then, the mission was clear, but today it is not. Over the past fifty years, the agency has been modified into a fragmented system unable to develop policies related to new environmental protection missions in the 21st century.[19] As with the UNEP, there have been calls for the EPA to be "reinvented" to meet the mission of the 21st century, without details on what this means. Like the UNEP, the EPA is saddled with bureaucracy, red tape, and even litigation. Its approach to environmental policy is more a function of the legal industry than of science and the supply chain.

Environmental nongovernmental organizations (E-NGOs) have increased significantly over the past decades. In the United States alone, there are nearly fifteen thousand nonprofit agencies related to the environment and animals, with total revenues of almost $20 billion.[20] This category of nonprofits was one of the fastest-growing sectors in the United States and worldwide from 2005 to 2015, with each of the twenty largest organizations having annual budgets of at least $1 million and some with budgets in the hundreds of millions.[21] For example, Greenpeace, headquartered in Amsterdam with offices worldwide, has twenty-four hundred full-time staffers and fifteen thousand volunteers worldwide, with a budget of approximately $40 million.

While some of the largest E-NGOs are very efficient with their donations, maximizing the amount that supports their cause versus what is spent on overhead—up to 90 percent percent—some are not, and more than 50 percent do not have a strategy.[22] With so many environmental policy organizations armed with lawyers and lobbyists going to war against large corporations and their teams, it's no wonder that the result is a lot of talk with little action. Advocacy groups are crucial to provide a voice to the voiceless—including marginalized communities, wildlife, and nature—but large investments should lead to more than just awareness campaigns. E-NGOs have a role in a 21st-century sustainability strategy, but environmental policy campaigns need to yield efficacious results that focus on continuous improvement and innovation in action, not words.

The CSR and ESG Revolution

CSR programs are all the rage. Approximately 90 percent of the top companies had a CSR program in 2019, up from 20 percent in 2011,[23] and 23 percent of *Fortune* 500 companies have tied their CSR programs to the UN's SDGs. However, only 0.2 percent of these companies have developed concrete methods and tools to evaluate the performance of their programs.[24] CSR programs are growing in importance within multinational corporations, supposedly as the result of calls from consumers.

A 2022 First Insight–University of Pennsylvania study found that more than 50 percent of consumers of all generations were willing to spend more for sustainable products and 90 percent of Gen Xers were willing to pay an extra 10 percent.[25] These studies fuel the rise of the prevalence of CSR programs in corporations, leading to an increase in the number of programs at universities and other educational systems that focus on the topic. Yet despite these efforts and the money spent on these programs, the results are less than adequate.

It makes sense for companies to have CSR programs in place because they are self-regulated, which provides an incentive to be a good steward voluntarily instead of facing mandatory environmental policies imposed by the government. In contrast to the Friedman Doctrine, which contends that a firm's social responsibility is only to increase its profitability, CSR programs consider the public an essential stakeholder in the business model. CSR programs need to focus on various critical public policy issues beyond consumer markets and shareholder wealth, which becomes tricky in terms of their effectiveness related to the environment. Diversity, equity, and inclusion policies in the United States and other developed countries have often taken precedence over environmental matters, which is a competing element of CSR corporate governance. And yet all these different elements of CSR policy are typically handled by a stand-alone department separate from the corporation's primary operations. Listen to a corporate earnings call for a *Fortune* 500 corporation, and you will hear the first half (or more) of the meeting focused on the financial performance of the company, followed by a discussion and questions related to typical "externalities" to the main business of the firm, according to its shareholders. For a soft drink company, for example, the financial performance talk might relate to keeping operational and material costs down to account for inflation, and the end of the call will feature a discussion on social matters and even future innovations such as new materials and recyclable plastic bottles. From the perspective of public corporations, which must report to their shareholders, the limitations of CSR programs are apparent.

ESG programs are even more the rage, turning into a battle within capitalism to make investments more sustainable or, from the other side, to sabotage future innovation. Over the last few decades, there has been a debate within financial markets about whether investments are evaluated based only on financial performance or also on environmental, social, and governance. In some estimates, a third of investment management assets are viewed from an ESG standpoint, but whether that's true and how performance is actually measured are less clear.[26] In 2022 and 2023, the allure of ESG programs in investment markets lost a bit of its shine, given that the top stock performers were oil and gas companies, a no-no for an ESG fund. As the United States and the world move increasingly toward an economic recession, focus areas are changing, an indication of the need for green and green, as emphasized in this book. Undoubtedly, financial markets need to lurch toward sustainable investments, such as solar and wind, and away from capital investments in the fossil fuels that will prevent us from hitting our 2050 objectives. But to do so, ESG funds and the consultancies profiting from ESG as "a thing" must clean up their act and hold themselves to greater accountability. If investors begin to feel these programs are more to the benefit of those administrating them than public policy, their effectiveness will follow a similar pattern of recycling programs that sound better on paper than in reality.

CSR and ESG programs overestimate consumers' and shareholders' willingness to put the planet ahead of the economy. Having spent a good deal of my career in consumer products, I have seen studies over the years that show an increase in consumer willingness to pay a higher price for a product if it is sustainable and a desire to fund investments in companies that produce these goods. While I have no reason to question the validity of the surveying process, the market data gathered by companies to signal consumer trends find little evidence to support these studies. There appears to be a significant amount of cognitive dissonance between what we know is ethical as citizens and how we behave in consumer and financial markets. Former US labor secretary Robert Reich considers CSR and ESG programs to be "BS" because he defines the present-day corporation as an entity that is neither greedy nor socially responsible—after all, its function is to manage a bundle of contracts, a transactional process.[27] In the present economic model, corporate managers—and even consumers, for that matter—care about society and the environment, but their job is to hit quarterly numbers and find the lowest prices, respectively. Corporations, consumers, and CSR and ESG advocates cannot develop and endorse these programs without reconciling these inconsistencies within the existing system.

What Should 21st-Century Policy Be?

Human beings have changed since 1846 when Thoreau journeyed into the woods and left the built environment behind. We have lost the natural association between ourselves and nature beyond perhaps taking the occasional hike or drive into the wilderness. As we try to identify a new relationship between the built and natural environments, with ourselves a part of both, we need to accept that Wilson's biophilia hypothesis is no longer possible; humans are no longer inextricably linked to nature through evolution. It is essential that our species continues to connect with nature, but we are no longer controlled by it—and moving forward, we must control it. Environmental variables such as viruses and our human nature remain, but we cannot isolate all eight billion of us in the woods. Perhaps separating from nature is the best way to save both it and ourselves, and I will discuss this as a critical solution in this book as a provocative but logical option.

We also must acknowledge that there is no evidence from our past to indicate that we can work together as one global community, despite our desire to do so. Humans possess no Gaia, the mother of earth. The UN does some good in the world, but not enough to make a material difference in the problems of CO_2 emissions and other environmental hazards. Advocacy for global and national agreements is founded more on faith than evidence. Intergovernmental and national agencies such as the UNEP and the EPA have the best intentions—but, as the saying goes, the road to hell is paved with them. Environmental policies devised in the 20th century cannot work to meet our needs today. These simple remedies solved water and air pollution in the United States, but today's challenges are more complex and nuanced as humankind's built environment grows in terms of both assistance to people experiencing poverty and harm to the planet. It's an overused phrase, but it is nonetheless true that we need to think globally but act locally—and action should take priority. History shows that throughout our evolution as a species, progress has been made through our actions as tribes, not as a global community.

NGO and E-NGO advocacy groups are essential in creating awareness of these problems. As is evident through the fifteen thousand E-NGOs in the United States, citizens are willing to fund these initiatives to support a specific cause. However, are so many disparate causes a practical element of environmental policy or more of a means to an end to the group and its mission? During the Trump administration in the United States, the policy was to roll back environmental policy as a diversion from the thousands of progressive policies stemming from these E-NGOs, leading to legal action and policy debates rather than science and solutions. Since the 1980s, the political playbook between these warring parties has been a distraction that impedes progress. Politicians and even NGOs, like any other institution, can be self-serving. Even if their work has intrinsic and financial meaning, it won't necessarily lead to meaningful solutions and progress. Today, we can see from the results that this is the case.

Perhaps the greatest misgiving of the public is CSR and ESG programs that are masked as market reform. Reich is correct in challenging them as a disfiguration relative to the definition of today's capitalist markets. Rather than the model of green and green that I advocate in this book, these programs profess to take the form of both, but without holding themselves accountable in doing so. Consumers and shareholders shouldn't simply read what's on the label of a product or the prospectus of an investment purchase, but demand to see results in the business and environmental case that was proposed to them. It's akin to every nation signing the Paris Agreement and, almost ten years later, not coming close to meeting their obligations. These are examples of how these large institutions are failing.

I favor environmental policy as one of the three key stakeholders, alongside science and supply chain, in meeting our 2050 sustainability objectives. However, it must support science and the supply chain more than driving the bus, as it has attempted for the past fifty years. I am not in favor of a government administration that egregiously rolls back critical environmental protections in an attempt to curry favor with large businesses, just as I am not in support of the environmentalists who wish to shut down all companies in a naive attempt to "save the planet." We need fewer red lights and more green lights, and quickly! Environmental policy in the 21st century must be focused on a different agenda. Rather than legal actions and policies directed toward large institutions, the approach must be to fund and support community and individual efforts of enterprise and innovation. It may also be based on policies that separate humans from nature, a practical approach to enable economic growth for humans while making room on the planet for ecological healing.

These are challenging and ambitious strategies to link the myriad local communities within a global network, but it is necessary. It's time for environmental policy to grow beyond simple advocacy and toward actionable solutions that take into account how humans get things done: through individuals and local communities. Later in this book, I will present a blueprint for these strategies.

A 21st-century strategy for environmental policy should be based on who we are as humans rather than who we would like to be. We shouldn't tie our future goals to the hope of 195 nations working together or multinational corporations, which are responsible primarily to their shareholders above others, considering social factors in their financial models. Both these scenarios are the world that we wish to live in, but we do not live in that world. In the world we live in, effective environmental policy is possible only through solving problems through individuals and local communities, one at a time and networked collectively, rather than through a global community or multinational corporate approach.

How would such a model be possible? I will discuss this in the remainder of this book. But first, before I discuss how a 21st-century approach to environmental policy can lead to a successful approach to sustainable development, I need to discuss the proverbial moose on the table: the plight of the developing world, also referred to as the Global South. Too often, environmental policy has focused on the developed world rather than the over five billion inhabitants of the Global South.

Instead, we must address the realities of the poorest parts of the world as the foundation of an environmental policy, and I will do so in the next chapter.

Chapter 3
Capitalism X.0

Capitalism is blamed for many of the problems in our world today, including the environmental crisis. In its present form, capitalism has been a root cause of many problems, but it is also a source of many solutions, such as a significant reduction in poverty and higher living standards worldwide. When discussing what should happen next, the answer is not as easy as either abolishing capitalism or continuing with it in its current state. To help us understand an optimal state for capitalism, we should turn to its founder, philosopher and economist Adam Smith, considered the "father of modern capitalism." It might be surprising to learn that Smith intended the field to be more of a social science, with the aim of optimizing a balance between what's best for the private sector and individuals, and the public sector and the common good. Any association between Smith's writings and a so-called free market lacking any controls or principles toward the common good is a misapplication. None of us can know what an 18th-century philosopher economist would make of the high concentration of CO_2 emissions in the atmosphere in the 21st century. Still, it's safe to assume that Smith would view what's happening as an imbalance between private enterprise and the public good that needs to be addressed.

The economy during Smith's era can be best described as "mercantilism" or "merchant capitalism." In Western European countries such as the Netherlands and Britain, private financiers funded merchants and gained significant prosperity from cozy relationships with the monarchies. The most famous of these companies was the East India Company, a stock company formed to enable trade within India that was estimated at its peak to control 50 percent of the world's trade. It had a standing army and was granted a monopoly on tea sales to the Americas, leading to the 1773 Boston Tea Party that eventually led to the American Revolution. If today's version of unfettered capitalism seems imbalanced, consider that during Smith's time, the East India Company was callous toward one of the greatest famines of all time, the Bengal Famine of 1770, which led to the death of a quarter to a third of the Indian population. Seventy years later, the East India Company smuggled opium into China to offset its limited silver reserves to pay for Chinese tea imports.

Much of Smith's writing focused on opposition to this form of mercantilism, a fixed and privileged system with an overriding objective of wealth accumulation rather than overall economic growth. He railed against the hoarding and misuse of capital by kings and private merchants alike, viewing both as a threat to the common good of the people of Britain. Rather than deploying capital to balance economic growth and the public good closer to home in Britain, Smith viewed these state-sanctioned ventures in faraway lands as an economic misgiving, like what is happening today. Smith was against colonialism for moral and practical reasons, seeing it as a distraction from the problems of ordinary people at home.[1] Rather than advocating for an approach to capitalism for the sake of those of wealth, Smith believed the capital should be liberated, similar to how the Founding Fathers suggested the American colonists should be in the Declaration of Independence, published in 1776, the same year as Smith's *The Wealth of Nations*.

So, before environmentalists propose abolishing capitalism as the solution to our problems, they should understand how Smith's writings present potential solutions, not problems, to a future strategy. Smith wrote that "the oppression of the poor must establish the monopoly of the rich" and that "profit is always highest in the countries which are going fastest to ruin."[2] These words ring true today. Smith was a champion of markets, a concept different from how they are often identified today. He was the founder of an innovative interdisciplinary view of social science in the 18th century to enable individuals to pursue liberty through markets rather than a concentration of power for the wealthy, public or private. From a Smithian approach to markets, capitalism can solve problems by factoring in what's in the best interests of communities and individuals, in contrast to the bias of today's institutional capitalism toward wealth accumulation versus distribution. It may be hard for some to fathom since Smith is often associated with unfettered capitalism, but his writings are our best reference points related to solving environmental problems. Although he is often described as a saint by the free-market capitalists and a sinner by the environmentalists, Smith could have the answers we need today hiding in plain sight.

A balanced public-private approach focused on problem-solving through markets is needed in the Global North, which is experiencing a love-hate relationship with capitalism, and in the Global South, which has been abandoned by it. A Smithian approach is more pro market than wealth accumulation, with the former being about solving problems and the latter for those who own the capital. Through markets, capitalism has reduced extreme poverty by creating a better balance between supply and demand, an outcome Smith would endorse. To address societal and environmental challenges, markets must be redefined within this problem-solving paradigm, and not just seeking maximum revenue and profit. I will call this next version "Capitalism X.0," a neo-Smithian model that seeks growth and balances through markets. Therefore, to solve the problems of the environment, the primary goal should not be to reduce CO_2 emissions but rather to understand the misguided nature of our markets that is leading to these problems in the first place.

Smith also understood the limitations of humans in society, as noted in his book *The Theory of Moral Sentiments*. To Smith, the purpose of markets is to provide social norms, and trust should shape institutions.[3] From my research and that of many others, it is clear that trust is best established at a community level as a problem-solving mechanism that enables individuals. A community-based approach to markets that empowers individuals is consistent with Smith's conception of markets when he wrote of the issues in the 18th century that are comparable to the problems we are witnessing today. Concentration of power, capital, and wealth within large institutions has limited trust even as it improves scale, leading to lower costs and more comprehensive product availability.

The goal of the 21st century must be to scale sufficiently to meet the world's needs, but within human systems, such as communities, where trust is held best. Capitalism enabled in communities networked to each other through improvements in technology and reductions in tribalism reduces distrust.

It is no historical coincidence that less than fifty years after the publication of Smith's *The Wealth of Nations*, the economy began to grow as a result of fossil fuel proliferation. Since 1820, there has been a hundredfold increase in the world economy, following thousands of years of scarce growth.[4] During this same period, the rate of extreme poverty worldwide has fallen from nearly 84 percent in 1820 to less than 10 percent today.[5] Many economists credit Smith for changing the nature of the field of economics and economic policy toward a liberalist tradition focused on freedom and liberty. He predicted, correctly, that England could not retain America as a colony and that the latter would become powerful and prosperous as a nation in the future.[6] His view of capitalism is consistent with traditional American ideas of individuals, economics, and markets that should work hand in hand to serve what's best for society.

Have you ever wondered why we emphasize the importance of liberty and democracy related to politics but not necessarily related to markets? Smith saw a connection between the two in the 18th century, and we should do the same today. Rather than abandoning capitalism because of the negative impact it has had on democracy and the environment, we need to implement a new version of Smith's approach—Capitalism X.0—focusing on competition, liberty, and individual enablement, to address the economic and environmental challenges we face. To do this, we need to create a 21st-century model of a community-based market system that provides economic liberty for the individual: a high-tech "invisible hand" model.

New (Sustainability) Economics

A new model for free-market economics is necessary to address the problem of 21st-century sustainability. Over the past two hundred years, extreme poverty has fallen precipitously, yet the combined rate of extreme and relative poverty is not substantially better than the poverty levels of the past. Poverty has become less extreme but not less relevant. Political economics has not gotten more sophisticated; instead, the benefits of poverty reduction are primarily the result of fossil fuels as an input to economic growth rather than human systems. Economic inequality remains high and is a variable in solving our environmental challenges. Today's model of economic theory considers environmental sustainability and relative poverty as externalities, separate conversations. CSR and ESG programs address the privatized form of capitalism without connecting it to public policy questions. In November 2022, the COP27 climate conference convened at a beach resort in Sharm El-Sheikh, Egypt, a nation with approximately 70 percent of its population living below the upper-middle-income class poverty line of $5.50 a day. The fact that the leader of the host country of this global conference, Abdel Fattah el-Sisi, is one of the world's most ruthless dictators was somehow lost on the UN when organizing the event. The deliverables from this conference, if any, weren't tied to individual and community enablement. Likewise, although the conference was held in a country wracked by high poverty levels, few tangible associations were made between poverty and environmental challenges. This conference was simply another lost opportunity driven by the limitations of our human systems and the lack of an ability to solve problems as a global community.

Smith's brilliance was in understanding the interrelationship between politics, economics, and society. Today, I am proposing the importance of an interrelated strategy that factors policy, science, supply chain, and economics brought together by the human system. The role of business within these human systems is to solve problems through markets, as noted by Smith. If capitalism continues to favor market concentration and capital (prosperity) hoarding, the opportunity will be lost for this capital to flow to sustainable investment opportunities. According to Smith, the true measure of a nation's wealth isn't the size of the king's treasury but rather the wages of the poor. His writings continue to affirm this concept of balance and quality of economic activity. Yet proponents of Smith's market theories have challenged the idea of the social welfare of the poor as a misapplication of capitalism. The 20th-century

Brazilian archbishop Helder Camara is quoted as sayng, "When I feed the poor, I'm called a Saint. When I ask why there are poor, I am called a Communist."[7] If the current focus of capitalism, defined as free-market economics, is directed toward greater economic growth, defined through hoarding, market and wealth concentration, and inequality, it threatens liberty, democracy, prosperity, and the environment. Smith identified political freedom as tethered to economic democracy, not seeing one as viable without the other. In the 21st century, the fastest and largest growing economy in the past twenty years has been the People's Republic of China, a communist country, not a liberal democracy, with a growing rate of inequality. With China's rise now threatened by a variety of factors, including its own approach to political economics, do the two economic superpowers face a growing threat that is problematic to the planet? It is clear the two economic superpowers, China and the United States, have the most significant, profound impact on the environment and growing rates of income inequality. Can changes be made for a positive role in the future?

Smith wouldn't be alone in considering the current global economic system perverse; few today, even from differing ideologies, would disagree with this. Some cheered that, before COVID-19, extreme worldwide poverty levels had been cut in half since 2000, a historic accomplishment in human history. However, since COVID-19, the levels have increased by half a billion people, or 8 percent of the world's population.[8] Lowering poverty levels has been possible through the increased burning of fossil fuels, which is destroying the environment. A financial system that enables the choice of more significant economic growth by doing further damage to the environment and further perpetuating extreme poverty is dystopian. And yet few economists are raising the problem that the current system cannot escape this dilemma.

According to the UN's definition, the term "extreme poverty" describes those earning $1.90 a day, which has been adjusted to $2.15 a day according to a recent World Bank study that considers inflation driven by the war in Ukraine, among other factors. However, a broader definition of "poverty" accounts for the approximately 85 percent of the world's population that lives on less than $30 a day. According to a World Bank definition of the term "poverty," there are thirty-three low-income nations that have a poverty threshold of $2.15 a day, thirty-two lower-middle-income countries with a threshold of $3.20 a day, thirty-two upper-middle-income nations with a threshold of $5.50 a day, and twenty-nine high-income nations with a threshold of $21.70 a day.[9] Individuals living under this broader definition of the term face a less severe existential threat, perhaps not one of day-to-day survival, but they struggle constantly and cannot focus on anything other than meeting the basic needs of themselves and their families. Suppose an estimated 70 to 85 percent of the world's population struggles for day-to-day survival, despite the breathtaking improvements in the past decades relating to extreme poverty. Given these statistics, is it feasible for us to separate the challenges of the environment from the economy? Environmental policy studies and articles have condemned plans as being short-sighted and not focusing on the long term, but the problem is the opposite; a more significant and immediate focus for most of the population may be our best solution for the longer term.

The issue of poverty and well-being isn't just an issue related to the environment only in the developing world but also in the developed world, including the United States. According to the US Census, the poverty level for an individual in the United States is $56 a day, or $108 a day for a family of four, assuming an eight-hour workday.[10] According to a study conducted by USA Facts, Americans make up less than 5 percent of the world's population and account for 20 percent of the world's income, yet 11.4 percent of the US population lives in poverty under this definition.[11] This definition of poverty may be relative compared to the rest of the world, but it's still a struggle: 61 percent of Americans live paycheck to paycheck.[12] Do the high-level math of adding those in the world in extreme poverty, those in more stable poverty, those relative poverty (in wealthier nations), and those struggling even though not in poverty, and the total adds up to more than 90 percent of the world's poluation, despite the dramatic improvements made over the past decades. With these data in mind, the global gross domestic product (GDP) has grown exponentially since the early 19th century. Still, it has not been equally effective in solving societal problems and has worsened significantly relative to the environment. For all the fanfare associated with the gains made by capitalism to pull billions out of poverty, this progress has limitations that have spilled into environmental problems.

In 1776, Smith called out the kings and mercantilists who stifled growth and innovation for their self-interest. Today, we need to call out the large institutions that are also stifling growth, innovation, and prosperity for a more significant percentage of the population and the betterment of the environment. A new approach to capitalism—Capitalism X.0—focuses on the interests of individuals and communities rather than large public and private institutions. It's the same problem that Smith addressed 250 years ago.

Environmentalism for Rich People?

Today, we live in a world where an estimated 80 to 90 percent of the population is struggling with poverty or economic difficulty and can focus only on their necessities rather than improving the world. The data also show that wealth accumulation and distribution are concentrated, leaving an unhealthy situation between the 80-plus percent who struggle economically and the nearly 20 percent who do not. In 2018, *New York Times* writer Neil Gross asked whether "environmentalism is just for rich people"—a fair but uncomfortable question. Gross is referring to a 1995 research study by sociologist Ronald Inglehart that found through a survey of forty-three nations that the adoption of sophisticated environmental policies correlates with the countrys' economic affluence.[13] According to Inglehart, environmentalism is an ideology for those who don't have more immediate and acute necessities of concern. Stating that "environmentalism is only for rich people" isn't intended as pejorative toward those conscious of and making progress toward finding solutions to this existential environmental challenge. Instead, it should be used to consider the concerns of the remaining 70 to 90 percent of the world's population. Duke Law School professor Jedediah Britton-Purdy made a case for environmentalism as a social justice movement—like it used to be before—noting that care for communities and the earth belong together.[14] Unfortunately, the term "social justice movement" has become a lightning rod within politics that separated the tribes, rendering it useless for bringing people together. And yet, whatever one's ideology, nearly everyone agrees with environmental protection if it is paired with economic fairness. We must educate the general population that we cannot have one without the other.

It's clear to see how corrosive and combustive this model of environmentalism can be perceived by the rural poor in West Virginia, the inner-city youth of Baltimore, and the extreme poor in Asia and Africa. And while highly renowned and brilliant astrophysicist Neil deGrasse Tyson had good intentions in wanting to focus the population's attention on science to solve problems, his 2016 proclamation that he was to be on a mission to "make America smart again" after the 2016 election of Donald Trump was insensitive. When America's middle class is paying higher prices for gas than they can afford and views a transition to clean energy as the root cause of the problem, environmental literacy and reflection should be more empathetic toward economic matters. When Trump announced his third run for president in late 2022, he whipped up the crowd by stating, hyperbolically, that the opposition is more concerned about the environment and the challenges that will arise three hundred years from now than they are with the people in his audience.[15] Scientists should use emotion and empathy to match their intellectual intelligence in addressing economics related to the environment.

Even more explosive than economic contention within the United States is a growing struggle economically and environmentally between the Global North and the Global South. Environmentalism is often too focused on the Global North, even though the epicenter of the most significant challenges exists outside its sphere of influence. Noted Singaporean academic Kishore Mahbubani writes and speaks about the unwrapping of this so-called Westernization toward a more balanced view. Statistically, 60 percent of the world's population lives in Asia, 17 percent in Africa, 5.5 percent in South America, and 8.5 percent in Latin America, with Europe, the United States, Canada, and Australia together comprising only around 15 percent of the world's population. Before the 19th century, the global gross domestic product was flat but tilted toward the East. Since the Industrial Revolution, the West has been responsible for most of the world's economic growth despite representing a fraction of the population. However, in the 21st century, the rest of the world—including many of the world's most populous countries—is seeking the same economic prosperity today as in the affluent West.

We hear a great deal about the rise of China and the prosperity of Japan, South Korea, and Taiwan; however, India, Vietnam, Indonesia, and other Asian nations are set to rise as major economic powers, and it won't be long before African nations also rise. As these economies grow and begin to use more natural resources, including fossil fuels, a significant strain is being placed on the environment. However, the wealthiest nations are still responsible for the most significant percentage of carbon emissions. Since 1850, the United States has been responsible for 20 percent of carbon emissions, followed by China (11 percent), Russia (7 percent), Brazil (5 percent), and Indonesia (4 percent), with the last two related to CO_2 from their land.[16] Industrialized nations, primarily in the West, have been mainly responsible for CO_2 emissions and have seen a favorable net benefit compared to poorer countries, mainly in the East.

Based on the data, it's difficult to escape that the more prosperous nations in the West will need to carry the more significant burden in solving this problem and cannot expect poorer countries to do anything other than survive; however, we seem reluctant to do so. The trick is to make this happen without punishing those affluent populations who would disagree with compromising their economies. Surprisingly, a greater association between poverty and the environment hasn't been created as an effective strategy that can lead to solutions such as those that I will identify in this book. For example, Africa accounts for less than 3 percent of the world's energy-related emissions today. It will likely be the fastest-growing energy market in future decades, with tremendous potential for renewable energy. While there is much to learn in how to deploy such a strategy (which I will discuss in chapter 6), think of how all of Africa—and even a large percentage of the planet—could be powered through the use of a massive solar energy project in the Sahara Desert if transmission challenges can be addressed! The business-case math is that investing in renewable energy in Africa creates significant benefits in eliminating poverty, stabilizing the environment, and creating economic growth. Through sheer mass, the continent of Asia will be where the

21st-century sustainability battle will be fought, given the sheer number of inhabitants, population growth, economic development, and environmental destruction in the present and future. In Asia and Africa, the scale of the problem and the solution's potential is breathtaking. According to an *Atlas of Economic Complexity* study from 2022, the forecast for severe outbreaks of social unrest in Asia and Africa in the next twelve months was between twenty-five and one hundred for many countries, with notable flashpoints such as Egypt and Pakistan, the latter of which some predict will be a failed state.[17] And yet in neighboring India, the government plans to invest $500 billion in clean energy, as it is the world's most populous nation. Despite our tendency to focus on the Global North, these challenges of economics and environment in the Global South will define our collective future.

Globalization X.0?

According to American economist Jeffrey Sachs, the world has been global since the modern humans left Africa some seventy thousand years ago, with globalization changing from age to age.[18] In his book *The Ages of Globalization*, Sachs discusses the seven epochs of globalization, starting with the Paleolithic era and the global dispersal of clans, through the Neolithic farming period some ten thousand years ago, and continuing today in the 21st century with questions about whether there is or will be an international rule of law. Nearly thirty years ago, *New York Times* columnist Thomas Friedman published his famous work *The World Is Flat*, which also trumpeted this new era of globalization of a "flat world" of the 21st century. Rather than seven stages of globalization, Friedman offered three: Globalization 1.0, from 1492 to 1800; Globalization 2.0, from 1800 to 2000; and Globalization 3.0, from 2000 to the present.[19] Finally, before the COVID-19 pandemic, a fourth version of globalization was offered to us: Globalization 4.0, discussed at the World Economic Forum in Davos under the theme "Globalization 4.0: Shaping a New Architecture in the Age of the Fourth Industrial Revolution."[20] The driving force behind this view of globalization is the foundation of the Fourth Industrial Revolution based on new artificial intelligence technologies, robotics, and others. Now there are even different versions of a Globalization

5.0, defined in a variety of ways, such as China as the world's leader or a so-called Imagination Society (whatever that means).[21] These models demonstrate our inherent desire to relate to some definition of globalization more than a concept. Yet, regardless of these flashy definitions and theoretical structures, there's no evidence that globalization has, does, or ever will exist. As such, let's define a new model called "Globalization X.0."

If "collaboration" replaces "globalization," perhaps these authors are accurate in defining these eras. What has been defined as globalization has been a model of partial collaboration around joint objectives, an economic structure, laws or agreements, and similar concepts. However, when using the term to mean full global collaboration, there are not four, six, or eight stages of globalization, but zero. These economists are correct that advancements in globalization have led to improvements in socioeconomics and technology. Still, gains are only limited if 80 to 90 percent of the population lives in a state of daily crisis. In using Sachs's seven ages of globalization, beginning seventy thousand years ago and continuing today, for all but the last 150 of those years that included fossil fuels, most of the world's population lived in extreme poverty, and there was marginal economic growth. For every Leonardo Da Vinci and Sir Isaac Newton, there were hundreds of millions who lived, suffered, and died a short life.

In none of these definitions of globalization is there a discussion of 80 to 90 percent of the world's population living in relative poverty. Friedman predicted that an emerging level playing field would eventually liberate people from extreme poverty; his prophecy was realized to a limited extent, but not in the sense of a broader 21st-century definition of poverty and opportunity. Of the hundreds of millions of Asians pulled from extreme poverty, some are returning, and hundreds of millions never progressed any further and remain under the same or similar circumstances. Even my blue-collar relatives in Baltimore, who prospered from manufacturing in the early to mid-20th century, no longer do so today, having suffered from this concept of globalization. And today, the entire planet is facing the challenge of addressing climate change that will pit their economic well-being against policies to improve the environment—at least according to the way our policies work today. Theorists love to discuss the concept of globalization, but the data clearly show that it has delivered on some promises but is not as comprehensive as some authors have stated.

I'm proposing an alternative model that I'll call Globalization X.0. Rather than developing some continuum throughout history, Globalization X.0 is the acknowledgment that this worldwide community has always been a fable. It is a shift from framing challenges at a global level toward the value of the local community and networking as a world. At its best, the definition of globalization can be a cluster of communities, an arrangement of interconnectedness for mutual interest (much like the internet). Perhaps as a modern, digital version of the legendary Silk Road founded more than two thousand years ago linking China to Rome as it winds its way through India, Central Asia, and Persia to Europe, today's global model is an ever-changing interlocking system of economic interests through newer technologies. If Adam Smith could have conceptualized the internet, it is how I believe he would design a global economy. Major worldwide events such as the COVID-19 pandemic and the Russian invasion of Ukraine should not mean an end to globalization but rather an understanding that such events have always occurred and always will occur. In this Globalization X.0 model, technology can enable individuals and communities rather than the large institutional model known as mercantilism in Smith's era.

An unrealistic flat-world model of globalization (pre-X.0) is demonstrated, perhaps with the best intentions, through organizations such as the World Economic Forum (WEF), an international NGO of the world's top thinkers founded in 1971 by German globalist Klaus Schwab. According to the WEF's mission, the world is best managed and led through a collection of the largest multinational corporations, leading political leaders, and other notable thought leaders and industrialists. Membership in the WEF is by invitation only, a posture that has opened the door for many conspiracy theories relating to its "new world order." While wild conspiracy theories related to the WEF are unfounded distractions, there are questions about whether it is effective for solving the world's problems, given its track record. An increasing percentage of the world's population in developed and developing countries has lost faith in this model of globalization as the savior. These approaches to globalization, including the worldviews of notable leaders such as Sachs, Friedman, and Schwab and intergovernmental bodies such as the UN, represent a top-down approach that has led to miraculous reductions in extreme poverty and fantastic technological achievements spread across the planet, but it has always been an unsustainable long-term approach. A Globalization X.0 model as presented in this book, even just as an alternative, is our best hope for the future.

Glocalization, Not Globalization

Adam Smith would have found it somewhat ironic that Archbishop Camara was called a communist when he asked why there is poverty and praised when he did something about it. Yet such a view is logical when we consider that problems of poverty are more likely to be resolved at a community level rather than through broad, sweeping government programs. For example, while the People's Republic of China is a communist government by definition and has pulled millions of its people out of poverty through its programs, it has struggled with fulfilling its mission to provide income equality, as demonstrated in its higher levels than nonsocialist nations. The moral of the story is that actions undertaken at the community and individual levels have greater opportunities for success, not necessarily through either a centralized communist government or a liberal capitalist model, but rather through enterprising community activities.

Humans live in a built and natural environment where most of the population lives in insecurity within a global supply chain system; even US treasury secretary Janet Yellen suggested that globalization has become a "race to the bottom" that has become global capitalism at its worst.[22] Rather than using free enterprise, innovation, and entrepreneurship to solve problems of poverty and the environmental damage it causes, these systems have enabled the damage as a disfiguration of how markets are supposed to solve problems beyond a zero-sum mentality that Smith addressed. To paraphrase Smith, markets should be problem-solving mechanisms for society. In the 20th century, global life expectancy increased from thirty-one years in 1900 to seventy-one years today; in 1947, about half of the world's population was malnourished, compared to 11 percent today; and 77 percent of income growth of the poorest in the world's population was due to economic growth rather than income distribution.[23]

These are signs of progress but also indicate a job left undone. To economist Thomas Piketty, income inequality is an intentional motive of capitalism, but that is too simple and academic an argument. Rather than replacing capitalism, a purer form can focus on *glocalization*. This community-based, networked structure can liberate individuals from acting out of self-interest, leading to work toward the common good.[24] As an alternative to today's institutional public and private capitalism model, the focus is on micro- rather than macroeconomics. The design I'm proposing is a community-based model that is neither communism, socialism, nor corporate capitalism, none of which can address these economic and environmental challenges. This new model of glocalism will be possible over the next decade by enabling technologies that prevented it from being possible in the past or even the current state. It is the same model Smith would have designed if he understood 21st-century technologies.

This concept of glocalization must enable individuals and communities using technologies such as blockchain and 3D printing as public utilities (which will be detailed later in this book) in much the same way that during the Industrial Revolution electricity enabled factories, highways enabled truck transportation, and ocean ports enabled global supply chains of massive scale. Public utilities are essential to promote economic growth. While there are plenty of calls for improving our 20th-century supply chain infrastructure of highways, bridges, ports, and other essentials, a more significant opportunity exists in developing the infrastructure for 21st-century supply chains. This Globalization X.0 model connects the economic world not solely through large-scale institutional supply chain systems, but in using technologies like the internet to cluster innovators and entrepreneurs. The intention is not to replace the present-day model of globalization through large institutions but rather to provide an alternative to improve competition and opportunity. Additional competition and opportunity within local communities and among individuals is the optimal approach to addressing our environmental challenges.

Fixing Poverty to Fix the Environment?

Naomi Oreskes and Erik Conway's 2010 book *Merchants of Doubt* provides evidence that scientists and corporations have manufactured "plausible deniability" concerns related to corporate and public opinion and policy.[25] The stories of the past reveal how oil companies and other large industries sought to conceal the problem of climate change related to the oil and gas industry. A 1980 American Petroleum Institute report warned that fossil fuels at their current use would not yield noticeable results in 2005 but would be a global catastrophe by 2060, and yet the industry pushed forward with greater production.[26] Today, while fossil fuel companies make billions in profits, they claim to be seeking a balance between the public's desire to transition to green and keeping the economy growing. We blame multinational corporations for profiting from fossil fuels, but at the same time, we ask them to pump more from out of the ground to keep energy costs lower. World leaders fly on private jets to global climate conferences while lecturing the rest of us about our carbon footprint. Western Europe, often seen as the most sustainable region on the planet, faced the challenge of energy shortages due to the Russian invasion of Ukraine, and, as a result, prioritized their economies over the environment through a greater use of coal and other fossil fuels. Now they understand the dilemma those in Africa and Asia face regarding the choice between the environment and economy, which is why environmental policies have failed for decades.

After thirty-five years of failed policies directed at CO_2 concentration levels, ocean plastic, sea level rise, and other issues, we need to acknowledge the problem of global poverty without believing it can be eliminated through global solutions. The richest in the world must take responsibility for this problem, but not through blaming and shaming, an end to growth, or the old methods of science, supply chain, and capitalism discussed in this book. Instead, the fortunate must focus on economic growth to reduce poverty and stabilize the environment, not just for them or their nations, but for the entire planet. There needs to be less of a focus on the symptoms, such as CO_2 emissions, and more on the root cause problem, such as poverty and the limitations of our human systems.

The irony of a Globalization X.0 strategy is that despite a lack of evidence that the world can act as a global community, the fortunate of the world must focus on poverty as a global problem. This riddle can be solved by attacking poverty as a community-based rather than a global problem. By addressing poverty within a community, we can make more progress on the environmental crisis than when we develop global climate change policies. This appears counterintuitive at first, since the planet is Gaia, an interconnected, synergistic ecosystem. Undoubtedly, environmentalists will struggle with this paradigm shift as they dream of the day when the world comes together in fruitful harmony and works together for the betterment of the planet. Such folly has often come at the expense of those who are the most unfortunate, sometimes known as the surplus population or the overpopulation problem. When we look at the data, we can see a direct correlation between poverty and environmental damage. The wisest environmentalist will shift their focus from carbon emissions and plastic pollution and take notice and action in Africa, one of the few areas in the world with high concentrations of more severe poverty and growing population numbers. This new approach to policy won't change that humans are not intended to operate as a so-called global community, but it won't matter when we prioritize communities and solve them one by one.

Poverty reduction must happen through markets; fairer, individualized self-interests; and enabling of the community that leads to innovations that address environmental challenges. The economics of Adam Smith should lead to the deployment of capital with the most significant opportunity for the economy and society. Individual entrepreneurs and local communities support the natural rights of man, an important topic in the 18th century as it is today. The innovator within the community is the balanced solution for the economy and environment.

Finally, the main lesson learned from thirty-five years of failed climate policy is that public and private institutions cannot solve these global challenges without getting most of the world's population on board with the strategy. If you ask a poor or struggling family in the developing or developed world about the 2015 Paris Agreement, most would either not know about it or openly bristle at the lack of assistance it provides for their needs. Likewise, why should developing nations act toward climate policy when the twenty wealthiest countries in the world are not holding to their agreement? This is policy stagnation at its finest. However, if the conversation begins with a focus on the economy and innovation toward the environmental fix, most of the world's population will get behind it. Finding solutions to a problem must start with an attempt at understanding. After that, the crucial variables need to be addressed. The following chapters will discuss the two critical elements of a new globalization model focused on science as the inventor and the supply chain as the innovator.

Chapter 4

Science for the Planet

The greatest science fiction book in history was the first ever. In 1818, Mary Shelley published *Frankenstein; or, The Modern Prometheus*, the story of a created life form that got away from a scientist with a God complex. In the summer of 1816, Mary Godwin was traveling through Geneva with her husband-to-be, Percy Bysshe Shelley, and Lord Byron. Cooped up in their house during a miserable rainy season at the end of the Napoleonic Wars (with their crippling effect on European trade), as the legend goes, the three held a horror-story-writing competition. Mary spent days dreaming up a story of a young, ambitious scientist, Victor Frankenstein, who created life from parts without a sufficient understanding of what he created and in the end was unable to control the monster. When everything went wrong, Dr. Frankenstein exiled himself to the Arctic and told his story to an explorer, Captain Robert Walton, who also pushed the boundaries of scientific knowledge and personal ambition. This novel, written more than six decades before the Second Industrial Revolution, was a prescient tale of what happens when scientists play God but have only a human understanding of what they have created.

It's been more than two hundred years since Dr. Frankenstein first told us his story. As a child, I learned of the story through cartoons and breakfast cereal, and I dressed as "the monster" one Halloween, wearing a rubber mask. It wasn't until three decades later that I learned who the real monster was: the scientist who abused his knowledge of science. Although it looked scary, the creation was not a monster at all; at its beginning, it wanted nothing more than to be accepted by humans rather than treated differently. It was grotesque in appearance, with parts assembled from charnel houses and meat slaughterhouses, but it was more than the sum of its parts. Ultimately, the creation killed Dr. Frankenstein's best friend and bride after swearing revenge on humanity for the suffering it caused.

Two hundred years after the novel was written, science has become considerably more advanced, and humans hold nearly total dominion over the planet—a reason to believe we are God. Scientism led to the Industrial Revolution, which took technological innovations beyond human imagination. Our collective intelligence quotient (IQ) has extended beyond our emotional intelligence quotient (EQ); we lack the human systems to keep pace with what we create. In the novel, the mythical Frankenstein's monster terrorized parts of Germany; today's anthropogenic impact is across the planet. Even as this has been one of the most impactful stories over the past two hundred years, we understand the story while unable to control our compulsion to create what we cannot control.

Policy, Science, and Supply Chain

My definition of "sustainability" has three critical planks: environmental policy, covered in chapter 2; science, covered in this chapter; and supply chain, covered in the next chapter. Environmental policy is largely based on social science led by institutions to understand and manage the natural universe. It is a human dynamic driven by observation, emotion, self-interest, and other nonempirical means, such as the cognitive thought of philosophy, religion, culture, customs, and some science. The origin of policy is largely of these nonempirical means; imagine being alive before modern science had answers for why an earthquake or a volcanic eruption occurs. Without science, society has to settle for answers that can't be proven. Yet these "answers" were more than just an explanation of natural phenomena; they were also methods of acculturation, bringing together communities through stories and storytelling. Religions are philosophies based on nonempirical evidence passed on through storytelling and faith, possessing value to societies beyond the rules of life and explanations of the natural world. Today, even as science has uncovered answers to the natural world that were explained through stories of the past, religions and philosophies remain, given their importance to a culture. Modern-day science has become the most trusted source of reason in societies, but it doesn't have all of the answers and often doesn't bring communities together like religion can. When science tries to overreach, as Dr. Frankenstein did and as happens today, its authenticity should be questioned, not much different from religions of today.

Over thousands of years, humans have been on a scavenger hunt for artifacts and knowledge and, through this process, set up human systems built on culture, tradition, and philosophies. In the West, a modern definition of science was born during the Age of Enlightenment and the Scientific Revolution of the 16th and 17th centuries, through the works of Copernicus, Galileo, Francis Bacon, and Isaac Newton. The past four or five centuries have seen the most significant technological innovations and economic growth in human history, primarily attributed to science and the scientific method. Science has been the foundation of these technological advancements, but as was true with Dr. Frankenstein, our EQ isn't keeping pace with our IQ. When EQ, or our human systems, don't keep pace, we lose control, as with the application of inventions, such as fossil fuels through our modern-day supply chains without thinking through the repercussions to society and the environment. When our so-called knowledge is greater than our ability to understand it, humans demonstrate our limited capacities, despite these technological innovations.

The most significant difference between religion and science is that while the former is intended to provide the answers, the latter, when exercised properly, is designed to ask the questions. The scientific method is a process of continuous improvement, aspiring toward ultimate knowledge without ever reaching it. Yet according to philosopher Paul Feyerabend, who wrote *Science in a Free Society* in 1978, science has become a "new religion" rather than seeking a separation between "science and state."[1] Feyerabend criticized the designation of "scientific experts," which he believed were antithetical to a democratic method and led to biases and prejudices from the scientific community. As Feyerabend noted, once science replaces state-sponsored religion, it loses its legitimacy as an inquiry-based discipline. Under the scientific method, all findings are meant to be questioned, with the development of a falsifiable hypothesis through the constant questioning of conventional thinking. Science should be an endless quest for truth, never settled fact.

Unlike religion, science should not be "believed in" but rather "believed out of"—inquired to be replaced. When science requires faith, it has loses its credibility. Especially in Western cultures, science has become too much of a de facto replacement for religion in its role in our culture. Science should never be accultured like religion; it should always sit on the outside of society as a regulator for the truth. As humans seek to account for this role, they place themselves as almost having a Godlike governance over the planet; the best place for science is counterculture, the methodology for invention and solutions allowing existing culture and mores such as philosophy and religion to deal with our emotions. Over the past thirty-five years, environmental policy has mixed legitimate scientific evidence with emotion, which has led to a failure within our human systems.

The third of the three planks is the supply chain: the industrial implementation engine of the scientific process. Contrary to what many believe, supply chains aren't the same as trade routes that have been around for thousands of years; instead, they were discovered based on the foundations of the scientific method, science, and engineering. Without the understanding, discovery, and application of fossil fuels in the modern global economy that took form after World War II, supply chains as we define them today wouldn't exist. In contrast to the definition of the term within a business school model, supply chains are more science, technology, engineering, and math (STEM) based on how they drive innovation in economies. Think of science as the inventor of technologies and the supply chain as their implementor within societies. For example, the plastic polyethylene terephthalate (PET) bottle was invented in 1973, but it was the supply chain system that enabled the use of over a billion plastic water bottles daily worldwide. Through data analytics, process engineering, and other STEM principles, supply chain management has become the enabler for science that has driven economic growth and environmental damage. The relationship between these two fields, science and supply chain, is among the most unsung stories in the past two centuries. They work best together when science is the inventor and supply chain is the innovator of these solutions.

The relationship between these three planks is an interesting story. Policy in the form of religion, philosophy, and politics has existed for thousands of years and is embedded in our culture and society. In the past few hundred years, science and supply chain have driven exponential transformational progress that has raised living standards and changed society. Yet recently science and supply chain have been losing credibility through the institutions that represent them. Today, only 29 percent of Americans are confident in medical scientists acting in the public's best interests.[2] While a loss of trust in the medical field has grown in the United States over time, the COVID-19 pandemic was a tipping point, not necessarily related to the efficaciousness of the vaccine but to the policy and institutional practice of the scientific discipline that led to lower levels of physical and mental health capabilities during a time of crisis. Science as a discipline wasn't sitting on the outside of society as a truth arbiter, but rather a political matter of public policy that was one reason for a growing lack of trust. Beyond the COVID-19 vaccine, a falling percentage of Americans are willing to trust any vaccination despite their demonstrated efficacy over more than a century. Scientific discovery, long considered apolitical, has become a political flashpoint that divides the population rather than unites us. Science has become a part of policy and culture, and as a result of this and other examples, it has an impact on its credibility to solve major challenges such as climate change.

Part of the reason for this division are the changes within science itself, from "little science" to "big science." From its origins, little science is the application of the scientific method by individuals and communities working on research and experiments at colleges and universities, or even by themselves, unaffiliated with any organization. The discipline became more institutionalized within public (government) and private (supply chain) organizations as its successes grew. Due to the success of supply chains to implement scientific findings, I believe this is when the transition occurred from little to big science, becoming a market system ripped from the labs. An example of this process is the story of J. Robert Oppenheimer, the lead scientist of the atomic bomb project in Los Alamos during World War II, who famously said, "Now I am become Death, the destroyer of worlds."[3] In using this quote from Hindu scripture, Oppenheimer sees himself as the modern-day Dr. Frankenstein and struggles with his transition from a member of "little science" to now a function of the larger institutional practice. Afterward, like Dr. Frankenstein, he expressed regret, allegedly telling President Harry Truman in a meeting that he had "blood on his hands," a remark that infuriated the president, which led to Oppenheimer being blacklisted.[4] Science was becoming institutionalized, losing its sense of community of the little scientists on the outside. While it was perfectly logical for the scientific discipline to take this path of discovery and the implementation of technological innovation, its role as an independent arbiter of the truth was being lost.

Big science also led to industrial relationships in the discovery, processing, and scaling of fossil fuels without understanding the long-term impact on the environment that would cause the highest levels of CO_2 concentration in the atmosphere in millions of years. In the 19th and 20th centuries, there was a significant increase in the number and scale of chemicals developed and manufactured in Western Europe and the United States. Today, the ten largest chemical companies worldwide—with operations in Europe, the Americas, Asia, and the Middle East—each have annual revenues of over $20 billion. It is estimated that today there are approximately 350,000 human-made chemicals on the market, a mind-boggling total without sufficient control procedures in place to ensure these new materials are tested to be safe prior to supply chain implementation.[5]

Nathaniel Wyeth is an example of a first-generation "big scientist," taking a role at DuPont in 1936 as a field engineer and later discovering the variant of polyethylene terephthalate (PET) now used to make plastic water and soda bottles. He developed the material while experimenting with a plastic bottle to understand pressurized liquids. In the course of his experiments, he developed a polymer that is now used more than a billion times a day, even though it wasn't specifically designed to be safe within the environment in either composition or scale. Wyeth was the inventor and the supply chain system was the innovator, with society losing dominion in controlling its use within the environment. As science and supply chains have become institutionalized, the benefits and problems have scaled exponentially. Have we, in Oppenheimer's words, become death, the destroyer of worlds, while simultaneously the enabler of the built environment and consumer markets? The purpose of science has changed from being primarily responsible for understanding the world through thorough, testable processes to being the front-end engine of the global supply chain. As a supply chain professional, I am not against the relationship between these two disciplines, but rather how it has been institutionalized. In this book, I will develop a new, healthier relationship for both the economy and the environment.

The Pace of Knowledge

For thousands of years, our understanding of the laws of the universe was limited. Relying on philosophy and religion benefited societies/tribes (EQ) but did little for our advancement of applied knowledge (IQ). As the field took root, little science conducted experiments and was applied on a smaller scale; this led to success, and we wanted more in both variety and scale. Technological and structural advances changed the nature of science through supply chains, the innovation driver of economic growth. Big science became an institutionalized approach tethered to large public institutions (such as governments) and large private institutions (such as multinational corporations). Even institutions that traditionally supported little science, such as colleges and universities, have become tethered to the process, relying on grants from large corporations or governments to fund their programs. Today's market intention is larger and more complex, beyond the capabilities of little science, in its original purpose of understanding the universe. As the driver of innovation, supply chains sought science to enable markets, leading the field of science to become a more critical link within the economy. As the engine of industry, science may have lost its role as an independent arbiter in balancing the environment and economy. In the future, science must both play the role of the inventor of solutions that can be implemented by the supply chain and act as an independent arbiter of the truth.

These changes to the role of science and supply chain are driven by the pace and scale of change as expected from nations and societies. In 1982, futurist Buckminster Fuller theorized that in 1900 human knowledge had doubled every century; in 1945, every twenty-five years; in 1982, every twelve to thirteen months; and by 2020, it would double every twelve hours.[6] At the beginning of the 20th century, the pace of change was alarmingly rapid compared to the past, but relative to today, it is unbelievably slow. If we compare the pace of change within science and technology with our ability to keep up, we begin to understand the disconnection between our limited capabilities and the need for us to make it happen as inventors and innovators. In biological time, the human species is almost the same as when we evolved as a distinct species approximately two hundred thousand years ago. We are 99 percent identical to the chimpanzees we diverged from genetically approximately four to six million years ago.

Furthermore, for most of civilization, the pace of change was extraordinarily slow, as one generation lived much the same as the next. Today, not only is there a great deal of change different from generation to generation, it is becoming increasingly difficult for us to keep pace with changes from year to year. With the good comes the bad. Science and its institutions have lost their ability to keep pace in understanding designs and applications and their impact on society and the environment. As became evident through the COVID-19 crisis, science and supply chain fields are struggling to keep pace with innovation. Through these struggles to keep pace, science and supply chain within the present institutional structures are becoming discredited, as is displayed today in the media. And if science is no longer an objective function to balance factors such as the environment, what is?

Big Science, Little Science

Formal science is said to have begun in 1660 when the twelve fellows of the Royal Society first met to promote the advancement of science. Since its founding, the Royal Society has had only eight thousand members, including 280 Nobel laureates and some of the greatest names in science, such as Isaac Newton, Charles Darwin, Michael Faraday, Albert Einstein, and Stephen Hawking.[7] The organization's original motto makes a clear statement relative to the mission of science: "Take nobody's word for it," emphasizing inquiry and knowledge advancement. From its earliest days until World War II, the Royal Society was a governing body over what is classified as little science. During the early days of science, there was a world of discovery awaiting scientists, as the invention of the microscope enabled them to see the nearly invisible, the telescope to peer into the far away, and more accurate clocks and compasses to navigate. In the centuries since, scientists have been making remarkable discoveries that have changed our lives, but recently, some believe there are fewer opportunities for discovery and significantly more scientists trying to discover them. In 2020, economists from Stanford and MIT published a paper titled "Are Ideas Getting Harder to Find?" raising the question relating to the efficacy of scientific experimentation.[8]

More science is conducted today than in the past, and there is strong evidence that scientific discoveries are costing more money and taking longer to figure out. Before 1940, few Nobel prizes were awarded for work more than twenty years old, but since 1985, that has been the case for most of the awards.[9] John Horgan, who wrote the book *The End of Science* in 1996, claimed that science had discovered many significant topics and today is left to nibble around the edges rather than discover fundamental revelations.[10] Another perspective regarding the field of science relates more to its changing scope of work. Timed to coincide with President Dwight D. Eisenhower's famous 1961 farewell speech discussing the military-industrial complex, Oak Ridge Laboratory director Alvin Weinberg coined the term "big science," relating to the shifting of science away from the "knowledge for knowledge's sake" that happened in the university system and toward larger, federally funded projects relating to national defense and other categories.[11] Shortly after that, Derek J. de Solla Price wrote his famous 1963 book *Little Science, Big Science*, which discusses the progression of the scientist from an individual who worked as a scholar at a university focusing on discovery to an administrative type who focused on writing grant proposals and publishing articles to achieve a level of measured credibility through journal rankings.[12] There is a correlation between the timing of the transformation of science and the institution of the supply chain system after World War II.

For the right reasons, big science became the invention arm of global supply chains, the innovator and implementer. This institutional approach led to the changing nature of science as an international competition, particularly between the United States and China, further evidence of the tribalism of progress. A 2022 report by Japan's National Institute of Science and Technology Policy found that for the first time China published the highest number of scientific papers—27.2 percent of the top 1 percent of cited articles—while the United States published 24.9 percent.[13] This research is funded by significant investments from public and private institutions, including focus areas of strategic initiatives, such as the 2022 US funding initiative of $200 billion in research and development in semiconductors, to be more competitive with China and Taiwan. In this case, science has become a geopolitical competition not just in making future computer chips but also in geopolitical lines, with Taiwan being a strategic interest for both nations.

The good news is there has been a significant increase in cooperation between adversaries in science and other areas such as healthcare. Collaborative research between the United States and China remains strong and is growing in some areas; between 2015 and 2020, the number of joint papers between US and Chinese scientists grew from 3,412 to 5,213, although the pace of growth is slowing.[14] Despite the rhetoric, US and Chinese scientists are continuing to collaborate. The bad news is that the situation has become adversarial in other research areas, particularly military defense and high-tech opportunities related to the environment. Science as an institution has made progress in collaboration, but it has also become a tribal weapon of competition. There are examples of cooperation across so-called global communities related to the challenges of climate change. The question is whether this is sufficient for the scale and scope of the worldwide challenges that must be addressed by 2050. Given the failures of global cooperation in even the recent past, such as during COVID-19, there should be limits to our optimism.

Compared to the past, few remarkable discoveries are produced by single-entity scientists conducting experiments in a lab separate from institutions. Scientists in the field, who often work for large institutions, are saddled with bureaucratic paperwork to get research grants, focusing on areas where the money is available rather than on a search for intellectual curiosity. PhD students, researchers, and professors must adhere to a "publish or perish" model, playing a game within a system that dampens curiosity and innovation. Simply applying for a research grant or submitting a paper to a top journal requires specific credentials, such as a terminal degree and closed-loop relationships, with those on the outside not even considered. Large national science research laboratories must deliver on the government's investments and direct their focus on where the money is. Those who challenge modern science as heavily invested in gaining research grant funding as their primary focus are not entirely wrong. Despite a legitimate emphasis on conducting science for science's sake, those in the field admit the process compromises them. The question we must ask is whether this institutional model is sufficient as the foundation to address our 21st-century sustainability problems. The honest answer to this question is that it is not; on the contrary, it serves an institutional purpose of self-interest and as an engine for the supply chain and economic system, not necessarily a bad thing. However, when basic research, or "knowledge for knowledge's sake," cannot be cost justified, and doesn't allow researchers to conduct interdisciplinary research as is necessary for the environmental crisis, the system is crippling the role of science relative to this existential problem.

Of the most significant concern is how these structures are impacting the work of science itself. Some research has shown that today's institutional system of science discourages scientists from taking risks. For example, a 2017 study that used a dataset of nearly one hundred thousand National Institutes of Health (NIH) grant applications found that reviewers seemed to favor ideas like their own experience.[15] Suppose the scientific process becomes so institutionalized that it discourages risk-taking and innovation in favor of confirmation bias. How can the most significant challenges we face, such as climate change, be addressed by such a system?

The End of the (Little) Scientist?

The stories of little science are the legends taught to students today. Marie Curie won the Nobel Prize for Physics in 1903 before being rejected for membership in the French Academy of Sciences in 1911. As a hobby adjacent to his role as a patent clerk in Switzerland, Albert Einstein developed the special theory of relativity, perhaps the most remarkable scientific revelation since Newton's Law of Gravity. Could either of these legendary scientists have succeeded in today's model? Despite his significant accomplishments and worldwide notoriety, Einstein rejected requests to participate in big science such as the US Manhattan Project. Einstein was a pacifist, but as a German scientist, he knew of the possibility that the Nazis would develop an atomic bomb, so he sent a letter to President Franklin D. Roosevelt that is thought to have influenced Roosevelt in starting the US program. Shortly before his death, Einstein lamented that his only great mistake was sending the letter to Roosevelt.[16] Einstein and Oppenheimer may be the last of the little scientists, yet their careers intertwined with the needs and requirements of the large institutions of the day. These two Jewish scientists wanted to serve their country during a war against an enemy that was trying to annihilate them. Still, they were conflicted in a scientific development beyond their capacity, given their high IQ and EQ. As credentialed scientists, they feared the use of a scientific application without a better understanding of it. Still, military leaders didn't have time to consider these implications, given the nature of their responsibilities.

In Price's work *Little Science, Big Science*, he notes that before World War II, the cost of science per GDP didn't increase, but afterward it did so exponentially, leading to an increase in the number of researchers and thus increasing the cost of science.[17] Because science is so expensive, its focus is almost entirely related to big science, as little science is cost prohibitive. The cost of technology such as an advanced electron microscope or lab conditions to isolate a molecule for a chemical is prohibitive beyond the funding from a large corporation, consulting firm, government research laboratory, or college or university. As a result, there is little opportunity for individuals like Einstein to ponder big ideas like his special theory of relativity. According to Einstein, his theory began as a thought experiment at age sixteen, and he persisted over eight years through extraordinary scientific insight, determination, and hard work.[18] Even beyond the likelihood that another scientist with the sheer intellectual of Einstein would come along, it's an open question whether that individual could navigate the institutional nature of today's science to work through such a momentous achievement without being sidetracked by a focus on grant proposals and journal rankings. As such, the possibility of becoming a relevant little scientist is undoubtedly dying, if not already dead.

There should be no doubt that big science is of critical importance today from a national security and global collaboration perspective; the US government's Advanced Research Projects Agency Network (ARPANET) created the internet in 1983. The Human Genome Project led to the mapping of human DNA, primarily completed in 2003 and fully mapped by 2022. The Hubble Space Telescope, launched in 1990, has dramatically improved our knowledge of the universe. These large projects have been critical in advancing science. A similar large-scale, international focus on the climate and other manufactured causes would be a significant movement in the right direction. Yet something is missing at the heart of science in this institutionalized approach to the field. Is this the end of science as it has been known to have existed since its onset in the 17th century? Is our present model, that of big science, something different from what was taught in my K–12 education about Newton, Curie, and Einstein? Will the future continue down this path or even take a more ominous turn, such as in the use—or misuse—of artificial intelligence (AI) in science? What the future holds is anyone's guess, but it will undoubtedly be different. The question is, what will a different model of science mean to sustainability in the 21st century?

The Future of (Open) Science?

In this book, I am making the case that solving environmental crises requires an integrated discovery and application approach. Part of the future of science, much akin to understanding ourselves, is accepting its limitations and incorporating other fields of thought, such as the social sciences, into finding solutions to significant challenges like climate change. Science will succeed in the 21st century as an inventor of technologies required to address challenges, but without today's God complex. Iranian philosopher Seyyed Hossein Nasr notes that the rise of human secularism gave us a Promethean ambition to dominate nature and turn it into a lifeless mass, a machine.[19] Nasr's book *Religion and the New Order of Nature* identifies the origins of today's science with Eastern and Middle Eastern traditions but shifts toward reductionism, anthropocentrism, and ideology from scientific investigation. Too often in the West, physical science, the social sciences, and philosophy (including religion) have become a competition over who owns the truth. The answer, of course, is none of them, when we follow the official approach of science. From the East, there is more nuance in the relationships, which some in the West don't consider to be science. In the West, the bastion of industrialism, the focus of science has been on the reductionism of problems and opportunities, the breaking down of an equation until it is understood. As such, the planet's natural resources have become compartmentalized as inputs in a production model. In Nasr's words, it has reduced nature to a pile of things, of machines, and as a result, we have lost our relationship with nature and other nonmaterial aspects of life. Has science, in its Promethean vision of the world, lost connection to it in a way that is impossible to control? Today, the pace of change is so alarmingly quick, and the planet as a basket of natural resources for us to use; we are like Dr. Frankenstein wandering through a charnel house for human parts. Is the field of science sufficiently mature to solve a 21st-century problem?

For science to be a complete and vital participant in addressing these climate and other anthropogenic challenges, it must be transformed from reductionistic and institutionalized to more nuanced, seeing beyond atoms and chemicals and fostering human civilization and its relationships. There is no doubt that science will and should continue to search for grand discoveries funded through big science, but the application of these problems may fall through cracks that could best be channeled to innovative, community-based science projects.

The future of science in shaping a balance between sustainability and economics must be supplemented to an open-sourced, community-based systems a 21st-century version of little science. This new model of science isn't intended to replace big science, especially given how important it will be in addressing some of the most significant challenges we face in developing sustainable technologies, such as renewable energy sources, a replacement for plastic, batteries to better store energy, and so forth. Instead, this new system of open-sourced science can become a community-based cluster of networks worldwide to unify a multidisciplinary approach to connect the dots and complement big science. This new version of little science can be the glue that unifies different applications of critical technologies and extends science beyond scientism, including the need for science to better balance economics and the environment.

It's not so crazy to conceive of little science being enabled through community clusters to solve significant problems. Consider today's science as the invention and this new unifying version as the innovation engine, a methodology of unifying ideas and applications into solutions. The issue in implementing an open-source, community-based model for little science won't be in limitations around making it possible as much as it will be in the dominant institutions limiting its possibilities. As an analogy, consider that some governments worldwide have restricted internet access to their citizens, preventing them from communicating and collaborating with others across the globe. Likewise, the dominant scientific community is a closed system that allows only those with proper credentials, such as a terminal degree, and from specific organizations to participate in legitimate scientific research. Science is also strictly defined within parameters, reducing the study of a problem within these discrete, specialized limitations. Ironically, some of science's leading researchers are the first to challenge the closed, reductionist nature of nations, suggesting that these policies are restricting progress, yet they practice the same principles based on their definitions of merit.

A 21st-century model of little science must be born to offset these institutional limitations. Open science might be the 21st-century version of little science, a movement to bring transparency to the scientific community and democratize it to solve global problems such as climate change. Today, science has evolved to a definition of a community that is united across its professional ranks, often requiring a terminal PhD degree and restricting the participation of others, and predominantly focused on institutional research for the public and private sectors. With the development of the internet in the late 20th and early 21st centuries, open access to analysis began to grow, leading to protests from established publishers. Peer-reviewed research is becoming more widely available but remains primarily closed.

The good news is that an open-source—if not an open-science—movement appears to be heading in the right direction. In 2011, the Research Works Act was introduced in the US House of Representatives. This bill would have restricted open-access mandates for federally funded research, but wasn't enacted. A similar movement has occurred in Europe toward more of an open-science platform, at least in terms of sharing findings from established, peer-reviewed research. However, this simply means publishing research funded by institutions rather than supporting open-sourced, collaborative research.

Open access to research is a good starting point, but it's not the same as open-source, collaborative research. For example, the Global Initiative on Sharing Avian Influenza Data (GISAID) is an open-source repository of genomic data of influenza viruses that started in 2008. This repository, which has more than 260,000 viral whole-genome sequences from 142 countries, was called a "game-changer" by the chief scientist of the World Health Organization (WHO) as the primary data source for the SARS-CoV-2 virus.[20] This is an example of how the sharing of research can lead to transformational outcomes, including the development of life-saving vaccines crucial to fighting a global pandemic.

One adjacent opportunity to fight the challenges of human-made materials in the environment is the Materials Genome Initiative (MGI), started in 2011 by the US National Science and Technology Council (NSTC) to address the need to design new materials for the 21st century. The first step in this initiative—similar to mapping the human genome, but a lot more complex given the number of natural and artificial materials in the world—was gathering and managing data to advance material science, as genomics does for healthcare. In November 2021, the NSTC introduced a new five-year plan for the MGI to unify the materials innovation infrastructure; harness the power of materials data; and educate, train, and connect the materials research and development (R&D)

workforce for the future.[21] If successful, the strategic plan will implement the technical infrastructure, enable the metadata and data required to invent and innovate new materials, and build an educational structure within the United States to make it happen. If this initiative achieves its purpose, it could lay the foundation for an open-source R&D platform for 21st-century materials and energy sources required to balance the economy and the environment. This topic of an open-source MGI will be important in the strategies to meet 2050 sustainability targets, and I will discuss it in detail later in the book.

Some believe there is sufficient evidence to demonstrate that research productivity is declining and harder to find.[22] Yet, ironically, the problems we face on the planet—such as existential threats from climate change and modern warfare—seem more extensive than before as they play out on a global scale. In the face of this paradox, there is the potential for human-centered science to assist in using AI to generate hypotheses, learn scientific rules and apply them, perform laboratory tests, and suggest new research proposals.[23] Yet a University of Massachusetts study found that training AI models to do natural language processing can produce a carbon footprint of five times the lifetime emissions of the average American car, the equivalent of three hundred round-trip flights between San Francisco and New York.[24] A strategy focused on effective institutional systems and futuristic technologies isn't necessarily the panacea to this problem, especially when such efforts may do as much harm as good.

The future of science and its impact on the environment in the 21st century must be more balanced toward little science—or, as I like to call it, community-based and networked solutions across the planet. Government initiatives like the MGI can be helpful when they extend beyond the technical and inventions to innovations, which must include the public. As I will discuss in the next chapter, the supply chain engine that drives the innovation of an invention must be transformed into a community-based system to balance the economy and the environment, as must the platform for science. Just as community-based supply chains will not replace today's long-tailed global systems, nor will little science's community-based platforms replace big science. However, just as long-tailed supply chains have been unable to solve the global challenges of the environmental and poverty crises, nor can science on a worldwide scale. Big science will be valid in some large projects like the International Space Station, the Hubble Space Telescope, and the CERN Large Hadron Collider, but for not climate change and poverty. Science must become more individualized, community based, and glocally networked to create innovations that community-based supply chains can launch into innovations. These sustainability challenges will have a greater probability for success when solved at local community levels versus as global aspirations. The approach must be more balanced between IQ and EQ.

Mary Shelley's *Frankenstein* demonstrates her enormous talent of creativity, but she could not have imagined the technological advancements that would develop to shape the 21st century. Yet Shelley was more prescient than many even today, three centuries later, in predicting the runaway impact science and technology can have on society. In our modern media, Shelley's novel is considered a science-fiction or horror story, but it is neither. Instead, Shelley's novel warns of what can happen when human information to achieve extends beyond knowledge. As noted by sociobiologist E. O. Wilson, "We [humans] have paleolithic emotions, medieval institutions, and God-like technology."[25] To solve this dilemma, Wilson called for science and society to run through synthesizers that put the right people together to use critical thinking to solve problems. Should not those "right people" be all of us, any of us, as individuals working in the best interests of our communities? A systems view of the world includes policy focused on people, science to enable invention, and the supply chain to implement the inventions on behalf of the people, all completed at the individual and community levels, networked glocally. Rather than science extended beyond our human limitations, this model seems more democratic, resilient, and problem-solving from a systems standpoint that includes society and the environment.

At the end of *Frankenstein* when the creation asks Victor Frankenstein to create a bride for him, it laments, "It is true, we [the creation and his bride] shall be monsters, cut off from all the world; but on that account, we shall be more attached to one another."[26] This quote demonstrates that the creation is wiser than its creator and has more emotional intelligence than the scientist. It is pointless to lament our disconnection from nature in the built environment. If anything, we need to connect differently, perhaps separate from the natural world, if we wish to save it. It may seem perverse to some to accept our creation of a built environment damaging the planet, but much like Frankenstein's creation, it is our best hope in solving the problem. Science and the environment can never return to what they used to be, which is okay if its intention is an approach to the future that is in the best interest of nature and society. The good news is that little science can be restored through technologies that will be discussed in the remainder of the book. But that's the easy part; the more complex challenge will be improvements within our human systems to create a balanced approach across the social and physical sciences in which we, as humans, must be essential elements alongside the natural world. We can no longer bifurcate IQ and EQ if we wish to achieve our 2050 objectives.

Chapter 5

The Supply Chain Paradox

Life in the 21st century is a paradox. Hundreds of millions worldwide can eat fresh fruit, vegetables, and seafood on demand, in any season, no matter where they live. Yet hundreds of millions face starvation, living a fate not much different from those in the Middle Ages. According to the United Nations Environment Programme, in 2019 approximately 17 percent of worldwide food production was wasted, despite close to 9 percent of the world's population going hungry.[1] The problem is likely much worse today after the COVID-19 pandemic and the impact on food production from the war in Ukraine. How is it possible that so much food is grown that it is wasted, and yet so many are suffering from hunger? To paraphrase E. O. Wilson, humans are drowning in information and starving for wisdom. How it is possible to have both such transformative supply chains and such desperate needs in the 21st century is a paradox. Some gaze to the heavens for answers to these perplexing questions, yet all our solutions are here on earth.

In many ways, human existence has always been a paradox, starting with an asteroid that hit the planet some 66 million years ago, killed the dinosaurs, and opened the door for us to evolve from small mammals. An asteroid 6 miles (10 kilometers) in diameter left a 110-mile (180-kilometer) hole in what is now the Yucatan Peninsula in Mexico, killing the dinosaurs with pollution caused by fossil fuels. However, new studies have modeled that the black carbon that polluted the air at the time of this mass-extinction event was from fossil fuels formed more than 100 million years earlier.[2] Therefore, contrary to legend, fossil fuels are made not from dead dinosaurs but from decomposing algae and plant life that died more than 200 million years ago. If this is true, then CO_2 emissions from burning fossil fuels were the cause of the death of the dinosaurs—which allowed mammals to evolve into humans but also, ironically, could also cause our destruction. Yet unlike the dinosaurs, who couldn't have controlled their fate, humans can address the peril facing us and other life.

"Political economics," as Adam Smith called capitalism, was of limited capacity before the advent of fossil fuels. Fossil fuels are the energy of the modern supply chain, a 20th-century invention stemming from the logistic feats of World War II. In the last chapter, I discussed the synergies between science and supply chains, and how the latter enabled the former to achieve exponential economic growth. Some erroneously believe that ancient trade routes, such as the Silk Road between China and Rome, were supply chains; however, these were merely trading routes where goods exchanged hands several times from trader to trader in a world economy that was exponentially smaller than it is today. The energy source for these trade routes was animal and human muscle, insignificant compared to the power required to move an ocean freighter as long as the Empire State Building is tall and that can carry more than twenty-thousand containers. For more than 99.9 percent of human history, we relied on remedial energy sources that led to nominal economic growth. Without fossil fuels, there would be no supply chains, and without supply chains, there would be limited globalization, economic development, and environmental challenges. Without supply chain, science would have never flourished as it has over the past centuries, and neither would the environmental challenges we face.

In this book, I present a case for three variables related to environmental sustainability: policy, science, and supply chain. Policy is the search for human knowledge through philosophy, religion, government, and some science. It focuses mainly on the social sciences, a general field less empirical than, but just as critical as, the physical sciences. The problem with relying on policy to solve the climate crisis has already been addressed in this book through two major problems: the limitation of humans to act as a global community, and the large percentage of the world population who live in some form of poverty and are primarily focused on their most immediate survival, not environmental challenges. As discussed in the previous chapter, formal science was founded a few centuries ago under the motto *Nullius in verba*, meaning, "Take nobody's word for it." In other words, structured, testable discovery methods are critical for developing theories of how problems in the natural world can be addressed, but these methods must be interdisciplinary to our societies. Yet these scientific methods led to the invention of technologies, such as the discovery and refinement of fossil fuels, but are, on their own, limited in their capacity to drive economic growth. When science originated in the 17th century, the global economy was estimated to be less than $1 trillion, compared to nearly $100 trillion today. Capitalism, a theory within the social science of economics, is often credited for much of this growth. Still, from its starting point in 1776 through the next few centuries, the economy grew only to around $1 trillion. It wasn't until the formation of the supply chain in the mid-20th century that the economy took off. Science and market capitalism enabled the supply chain that drove today's $100 trillion global economy, leading to a system that even its creator could not control, for better or for worse. It's the great supply chain paradox.

The Affluent Society

The Civil War (1861–1865) is the darkest era in US history, but in second place is the combination of the misery of the Great Depression (1929–1945) and World War II (1939–1945), which was four times longer. When the war ended in 1945, after sixteen long years of misery and death, there were fears from leading economists that another economic depression would follow. Two future Nobel Laureate economists, Paul Samuelson and Gunnar Myrdal, predicted that throwing 10 million men back into the economy would result in a severe depression, possibly leading to an "epidemic of violence."[3] At the end of World War II, the United States was a wartime economy, with 55 percent of its GDP built on government spending. However, even before the war ended, US businesses began to plan for peace by converting their manufacturing capacity from wartime to peacetime, taking advantage of a society with a high savings rate and pent-up consumer demand. In addition, the US economy expanded for many of its citizens through government programs, such as the GI Bill, which sent many service members to college, and the Federal Housing Administration (FHA), which provided loans that enabled suburban home ownership.

For the first time in modern economics, the primary focus was not on the means of production to eliminate scarcity of supply but instead on the induction of demand for economic growth through markets. When Smith wrote *The Wealth of Nations* in 1776, the economy differed significantly from that of the post–World War II era. Smith's development of the concept of capitalism used markets to create a multiplier effect to improve production for supply to achieve an equilibrium with demand, because the former was always less than the latter. After World War II, the role of the government was to convert utilization from a wartime model to enable economic growth through capital investments and consumer demand. Investing in public infrastructures—such as dams, highways, public utility systems, and higher education, among others—would enable consumer-driven economic growth. The US economy was the largest in the world due to its manufacturing capabilities. It was left undamaged, leading to unprecedented economic growth and prosperity for those allowed to participate.

Through the necessity in avoiding another Great Depression, the US economy shifted from wartime to peacetime consumer production, moving from an economy with the objective that demand enable economic growth through production to an economy focused primarily on the demand side and less on production strategy. As such, the economy was more transactional and less focused on a value proposition. In his 1958 book *The Affluent Society*, economist John Galbraith argues that an economy based on hedonistic consumption rather than meeting basic needs would lead to economic inequality. This prediction most certainly has come true. Further, Galbraith notes that in such an economy, "demand is not organic" and is primarily driven by the customer but through advertisers as "consumer-demand creation" for economic growth.[4] This "affluent society" approach to economics has led to a path to capitalism based not on the economic balance of supply and demand but on achieving growth through consumption and financial systems. In Smith's model, capital is strategically deployed to enhance further economic growth opportunities to balance supply and demand. In this demand-driven model, capital is deployed transactionally to enable production to grow by artificially inducing demand that is based not on need but on wants. Through the growth of demand came the growth of consumption and environmental damage.

In 1899, six decades before Galbraith expressed concerns about the consumer economy after World War II, economist Thorstein Veblen wrote *The Theory of the Leisure Class*, in which he coins the term "conspicuous consumption," a concept that is well understood in the 21st century. With the advent of the Industrial Revolution in the late 19th century, Veblen saw a shift away from economics as a framework of rational agents seeking to improve economic conditions as a balance of supply and demand toward an objective of social status and prestige.[5] As a result, capitalism shifted away from Smith's intention to reduce scarcity and enable balanced growth to become a system driven by conspicuous consumption, financialization, and the commodification of products and labor. Ultimately, the supply chain system would become the driving force of this "demand chain" model that optimizes capital, raw materials, and efficient processes and transportation on behalf of consumers and shareholders. In this model, investors increase their return on investment by increasing revenue and decreasing costs through greater market penetration, inducing sales through marketing and expansion into new markets. The supply chain system is the back-end enabler of marketing and advertising by unleashing efficiencies in production and distribution through technologies invented by science, such as fossil fuels, computers, and the internet. And, of course, this demand chain model isn't good for the environment.

Understanding the difference in the application of capitalism today compared with Smith's intentions is crucial in understanding what has become imbalanced in the US economy after World War II. Today, very few people in the United States can remember a time when capitalism focused on balancing supply and demand rather than primarily on consumption; as a result, many in the younger generation believe capitalism itself should be scrapped altogether. The concept of "millennial socialism" emphasizes how the younger generations—the millennials and Gen Z—are more supportive of socialism than of capitalism, given the latter's failures related to income inequality, the loss of the American Dream, and climate change. All the generations alive today, starting with the Baby Boomers, have lived during the era of this affluent society of the demand chain, creating an association that capitalism focuses only on excess, material wealth, inequality, and damage to the environment. In its present application, this is true, but to solve the everyday challenges of poverty and the environment, we need a reformation, not a replacement.

Scarcely anyone alive today understands life without the modern supply chain system. Imagine a time machine that takes you back to the 1800s in America to spend time with your ancestors on the frontier. You wake up at dawn, planting, chopping wood, hunting for food, and other similar activities. After a physical twelve-hour day, you sit by the fire with your ancestors, and they ask you what your life is like in the year 2024: how you get and make your food, clothing, shelter, and so on. Your response is alien to them; they don't understand what it means that you "swipe a card" and products are handed to you, and they are most surprised that you don't know the origins of these products. You mention that much of it comes from hundreds if not thousands of miles away, you think (but you aren't sure). Nearly all of us eat meat but have never killed an animal, and we have closets full of clothes but do not know how to sew.

It's only a recent phenomenon, over just a few generations and even less so in other parts of the world, that our lives have come to depend on something we do not understand: the supply chain system. Many of us didn't even realize what a supply chain was until it led to shortages during the COVID-19 pandemic. But despite calls from environmentalists for a return the good old days, very few of them would be content if they knew what life was like before these supply chains existed, as is suggested in this time machine analogy. And yet, despite what would be considered by all generations who preceded us as a "leisure society," income inequality and overconsumption have been perpetuated through a disconnected global supply chain system that is bad for the environment. As supply chains have become global, the environmental impact of our consumption has also become disconnected from us, a primary source of cognitive dissonance between our behaviors and beliefs. Maybe if our ancestors came back to the present day with us through this time machine, they would be impressed with some aspects of our lives but would probably expect more than what they saw.

The problems of the 21st-century supply chain stem from its successes, not its failures. I say this from experience, having spent decades running large operations that achieved high standards of efficiency, safety, quality, and service to consumer markets. Supply chain professionals are often the heroes in our modern-day society, enabling the distribution, for example, of the COVID-19 vaccine in such a transformational manner; delivering bananas, coffee, and other fresh, exotic products around the world at very affordable costs; and enabling the next-day delivery of online orders and other expected miracles daily. But what is essential to understand about the field is this: Supply chain professionals deliver near miracles daily, but these actions do not focus on a balanced supply-and-demand equation within a community or nation, just what's best for consumers and investors. Increasing CO_2 emissions is simply a symptom of a more significant problem. Externalities such as poverty, income inequality, and environmental damage must not be on the outside in the 21st century.

Our Efficiently Broken Supply Chains

To summarize, progress was advanced in human civilization through policy, science, and capitalism, but substantial economic growth did not happen until the advent of fossil fuels, which led to global supply chains. And yet most consumers understand little about how these systems practically run their lives until something goes wrong. As a long-time supply chain professional, I can relate to how our industry is a modern-day problem solver. Still, we solve the problems only as they are defined and consider all else as externalities. When I worked for a multinational consumer products company, my job relied on my ability to solve problems for my main stakeholders: our stockholders and consumers. Today, it is fashionable for multinational corporations to have a corporate social responsibility and environmental social governance programs. Still, until science, supply chains, and financial markets are reformed, these are simply tactics rather than substantive actions. The primary goal for today's supply chain professionals is to increase shareholder wealth through increases in revenue and profit through innovation. The model will only change once supply chain innovations can increase shareholder wealth and customer satisfaction while improving the environment. This should be our goal for the future of supply chains.

Today's supply chain systems are not directly incentivized to improve the environment. Numerous studies have shown that consumers are willing to pay more for sustainable brands when they shop. For example, a 2021 Global Sustainability Study found that 85 percent of consumers worldwide have shifted their focus toward being more sustainable, and a third are more willing to pay more for their products, but these numbers are not evident in consumer markets.[6] Many studies conducted over the past few decades have yielded similar results. Google the question, "Are consumers willing to pay more for sustainable products?" and you will see that virtually every study concludes that consumers are becoming more sustainably focused. However, if this were really true, don't you think that the large consumer products companies would provide their consumers what they want?

In the run-up to the 2022 midterm elections in the United States, an NPR/*PBS News Hour* poll found the top three issues to be inflation (30 percent), abortion (22 percent), and healthcare (13 percent), with the environment not even among the top seven issues.[7] And these surveys were conducted in the United States, one of the wealthiest nations in the world, not among the billions of people worldwide living in relative poverty. In today's model, financial markets are the ultimate judge, and these results emphasize that consumers and investors behave differently from what is noted in surveys. This is increasingly the case in today's economy of higher inflation and greater economic uncertainty. Because supply chains are good at what they do, they are exceptional at understanding customer demand signals regarding all market elements, specifically price. The best supply chain companies are data-driven; they gather, process, and utilize enormous amounts of consumer data to win in the marketplace. In a competitive marketplace, they must read these signals properly to win against their competition, as seen in the ruthless consumer market battle between Coca-Cola, PepsiCo, and Keurig Dr Pepper. The leaders of these companies want to do the right thing for their children and grandchildren, but they must focus on the markets relative to consumer demand and competition to stay gainfully employed. I once participated in a global task force in Washington, DC, focusing on the ocean plastic problem. One of the other members was a burly man who represented one of the largest oil companies. At first glance, many of the task force members, especially the nonprofits, stereotyped the man as someone who was at the session to serve an apologist for his company, but during the session, he broke out in tears regarding his company's challenges in addressing the environmental impacts of their products and the consumer market. The fact is that most consumers, who focus more on the price of oil and other products than their impact on the environment, are just as complicit as he is. There's no simple answer for companies and existing consumers within the supply chain to encounter climate change, ocean plastic, and the like, as they must focus on shareholder wealth and consumer demand. New markets and consumers must be developed. The public and special interest groups may complain about companies and their practices, but our activity as consumers is what sends the demand signals. Cognitive dissonance happens when consumers protest oil companies while complaining about higher gas prices. Our supply chains are efficiently broken and must be fixed to solve this disconnect.

In reality, most Americans and others from the Global North ignored the supply chains' efficiently broken nature until it directly impacted them. Growing up in Baltimore in the 1970s and 1980s, I spent nearly every Saturday with my grandmother, who lived in the blue-collar southeastern part of the city. I was there during the downturn of the city's industrial base, including the Bethlehem Steel factory in Sparrows Point, close to my grandmother's rowhouse. I didn't know what a supply chain was, but I did experience its impact on the community. I wasn't a great student in high school and college, and I had limited options. In an earlier era, I might have worked in manufacturing, like my mother's side of the family. Yet, given their experiences, I was terrified of manufacturing and the supply chain and wanted to move as far away as possible, so I ended up in the financial services sector. Many who lived in the city of Baltimore escaped while they could, and the rest have languished due to the broken, successful supply chain that enabled them so effectively as consumers but left them as workers. If you ask someone from industrial cities like Baltimore and Flint, Michigan, or any of the numerous rural towns in the United States, they will scoff at the notion that these supply chains started to fail in 2020; they have been afflicted by deindustrialization for decades, even while the rest of the nation moved forward.

I did eventually find my way into the manufacturing and supply chain field, working for one of the largest beer manufacturers in the United States. Fifteen years ago, I began to write and publish about these challenges with our broken, successful supply chains at a time when few were paying attention since the supply chain worked just fine for the white-collar audiences that read my publications. Of course, politicians have been railing about the failures of the manufacturing sector and how jobs need to be brought back—but this is mere pandering for votes without sufficient plans for how to do so. The fact is, the present-state supply chain system is transactional. Because labor costs in the United States are much higher than they are elsewhere and the advent of information and transportation innovation have made it possible to produce goods great distances away from their target consumers, it was never a serious debate; supply chains are optimized at the global rather than the community level, focusing on cost versus value. As a result, the environment is negatively impacted by commodifying consumption and investment for the sake of workers and communities.

After World War II, supply chains were the foundation of economic growth in the United States, then they turned global, leading to higher income inequality and environmental disruption. We live in a built environment where production and consumption are separated, and events happening in one region of the world impact others. Governments, communities, and individuals lose agency when a manufacturing plant is moved from where they live to somewhere far away. From a corporate transactional standpoint, this is considered progress—and it is, for you as a consumer and an investor, but not necessarily for you as a worker and a citizen. An efficiently broken model of a global community is evident in the fluid supply chain systems that move raw materials, components, and finished goods worldwide. For example, your smartphone has raw materials and components sourced from dozens of nations and was assembled in Asia. Consumers are unaware that some of the foods they eat or the medicines they take are made overseas, often in the Global South, where workers' wages are significantly lower than they are in the Global North. Environmental standards are lower, too, and the transport of product across oceans in massive steamships is hurting the environment—but most of us know nothing about it. Problems cannot be solved if they aren't understood, and while consumers are beginning to understand that automobiles emit carbon dioxide into the air and your soda bottle is unlikely to be recycled, few understand the magnitude of today's global supply chains driven by a consumer economy. If more people understood these implications, such as the detrimental impact on communities and the environment when the jobs go overseas, perhaps the public would ask for necessary changes in the supply chain. Our lack of an understanding of such important elements of our daily lives has become a detriment to us and the environment.

Community-Based Supply Chains

In 2005, when Indian economist Raghuram Rajan presented a paper titled "Has Financial Development Made the World Riskier?" at a banquet in the US honoring outgoing Federal Reserve chairman Alan Greenspan, he was criticized by leading economists for being reckless and alarmist. A few years later, during the 2007–2008 financial crisis, Rajan was vindicated as a clairvoyant regarding his concerns about the stability of an economy driven by the financial sector and its casino mentality. Yet Rajan was hardly a clairvoyant, as the data clearly showed that the economy's weaknesses were driven by financial markets. In his 2020 book *The Third Pillar*, Rajan notes how the struggle between the two large institutional pillars discussed in this book—the markets and the state—have led to negative consequences related to the third pillar, the community.[8]

Macroeconomists, policymakers, and executives of large multinational corporations have focused on these two large forces without considering their impact on the local community. In terms of the business model, a multinational corporation functions within a global supply chain system to optimize revenue and lower production costs, with its impact on the local community as an externality. Before I knew about supply chains, I learned this lesson when industrial jobs left my hometown of Baltimore without anything to replace them. Many fled the city and rural towns, and what remained for those who couldn't go were mostly service jobs and the illegal drug trade. The large multinational corporations were doing what they were designed to do: meet their fiduciary responsibility to their shareholders, who don't necessarily reside in a corporation's city or nation of origin. In meeting this responsibility, they may find it optimal to outsource production to workers making a dollar an hour, often working without safety and environmental standards, through innovations such as the internet and efficient ocean shipping and logistics. The existing supply chain model is a long-tailed, globalized system that has led to spiky globalization rather than a flat world. These companies and the professionals who work in them aren't necessarily evil, as they are sometimes labeled. Rather than a giant effort to redesign and reform these existing institutions, an effort that would take too long and is unlikely to succeed, the goal should be to provide an alternative: the community-based supply chain system.

In my last book, *Reinventing the Supply Chain: A 21st-Century Covenant with America*, I defined the solution to this problem as a community-based supply chain system based on emerging technologies and improved processes.[9] This new supply chain management model for the 21st century focuses on a glocalized, networked model that enables economic growth within the local community, not as a replacement for today's long-tailed, globalized systems but rather as an alternative to them. The key to the success of this new community-based system is an investment in emerging technologies by the public sector and the distribution of them like that of public utilities. Technologies such as blockchain and 3D printing will be crucial in the future of the supply chain system. Still, suppose these are implemented only within the private enterprise sector maintained by the largest multinational corporations. This would lead to the same economic and environmental problems we see today. According to Rajan, economic activity must become more localized to benefit the community. When it is not, societies encounter the issues we see today, including the disenchantment with society that leads to extremist groups and partisan politics. Rather than moving from an economy driven by one large institution (multinational corporations) to one driven by the other (national governments), the focus needs to be on the third pillar, as Rajan calls it—or, in my supply chain model, the community-based supply chain. The good news is that these emerging technologies make it possible to consider how to make this a reality within communities that have been hardest hit in the United States and elsewhere, such as my hometown of Baltimore.

Despite the promise of implementing 21st-century technologies to fix these supply chain challenges, there is a vast and widening gap between the haves and have-nots within the United States and around the world. These so-called digital divides exist across nations and within nations, as well from community to community. To address the human-made environmental challenges discussed in this book, there must be a focus on connecting economics, poverty, and the environment to establish a future strategy. To do this, not only must the model be focused on communities within a public utility model rather than only privatized, but all within the community must have an equal opportunity to succeed within this 21st-century supply chain system. Without an acknowledgment of the problem and a focus on education to address the widening digital divide, the issues will persist and even worsen.

Here is the good news: a community-based supply chain model is the definition of capitalism focused on deploying resources to solve problems, as outlined by Adam Smith. Today's model of capitalism is focused on optimizing revenue and cost transactions for the benefit of the consumer and investor across the planet, without regard for value. When well-intended environmentalists suggest that growth is the problem, they conflate the potential of markets, as devised by Smith, with how they are misapplied today. Healthy growth, seeded in local communities, is balanced growth; unhealthy growth is where some communities consume but do not produce and some are low waged and cannot consume. Arguing that capitalism, along with its supply chains, should be replaced and growth ended demonstrates a lack of understanding of the realities of the world's economy and societies. Community-based supply chains are more possible than ever, able to be enabled through emerging technologies of blockchain, networking, 3D printing, and AI.

Growth for the Environment

A primary reason why environmental policy has been ignored since its inception is its often myopic stand that economic growth is the problem. Environmentalists who propose "an end to growth" are focused on maximizing a benefit to the environment at the sake of the economy, a conditional argument that sounds better in concept than it is in practice. Rather than an either-or model, a new system must take a both-and approach that focuses on the growth needed for the global economy, down to its communities, in a way that isn't harmful to the environment. After thirty-five years of this false narrative of green or green, it's time to reinvent today's supply chains to bring these goals together.

In the United States and other nations, the level of trust in these public and private institutions is at an all-time low, and for good reason. Rajan notes that most people identify with their communities and as individuals (the third pillar), and these entities are not enabled within today's long-tailed global supply chains.[10] Not only are people facing greater alienation and uncertainty in how they live their lives in the short term, but there's also a looming sense of fear about the longer-term future related to existential matters such as the climate and the environment. Supply chains have led to the most remarkable economic transformation in human history, yet they are at an inflection point as they fail us. For some, these failures happened sooner, even decades ago; others are facing these issues starting just recently through the COVID-19 pandemic.

In the remaining chapters of this book, I will focus on a 21st-century model to balance policy, science, and supply chain to enable economic growth and environmental sustainability. Solutions focused on energy, food, water, materials, and our human systems are what is needed to achieve necessary environmental goals by 2050. Is it possible to tackle all these goals rather than just what is always the focus area, energy? The answer to this question is that we must, because they are interrelated to the goal. For example, if we focus solely on energy without understanding how the food and material production processes are dependent on it, what have we accomplished? Part of the problem of a lack of progress in meeting these future goals is a lack of details in a plan.

Perhaps more so than in other fields, supply chain professionals are confident they can solve a problem when there are clear objectives related to defining it. Today's supply chain leaders aren't focused on environmental challenges because they are so focused on financial markets. When we bring together the goals of financial markets, science, supply chain, and the environment into a unified strategy, we will create new systems focused on these challenges. In the remainder of this book, you will understand how solving these environmental challenges isn't as insurmountable as often described, when we focus on understanding the root causes and creating solutions for them. In this case, the root-cause problems are a lack of any evidence that problems can be solved as a global community and the problem of relative poverty in the world. Taking these into consideration and allowing science to invent and the supply chain to innovate will lead to success, as I will discuss in the remainder of the book.

Chapter 6

The Future of Energy Is Bright (and Breezy)

I am optimistic that with the proper focus on policy, science, and supply chain, we can address the climate challenges we are facing by 2050. To achieve these objectives, we need to understand and agree on the most basic facts of the problems we are trying to solve, such as the CO_2 emissions problem. CO_2 levels have risen dangerously high: above 400 parts per million, a rate not seen in millions of years. Before these anthropogenic changes, the planet was stable within the 170–300 parts per million range for the past eight hundred thousand years.[1] According to some, such as Gregory Wrightstone, executive director of the CO_2 Coalition, CO_2 levels have needed to rise rather than fall given that they have been as low as 182 ppm over the past 140 million years, dangerously low to sustain plant and animal life.[2] Low CO_2 levels of 150–200 parts per million can threaten plant life, and higher levels, even over 1,000, can lead to enhanced photosynthesis. However, the framing of this argument is misleading given that humans have been around for only two hundred thousand years; not tens of millions of years ago, levels were much lower; and higher levels of CO_2 also lead to a loss of nutrition and drought, which impedes photosynthesis.[3] To solve these climate challenges, we need rational conversations of problems to develop appropriate solutions.

Carbon dioxide is a tricky gas that fits within the Goldilocks principle on earth: too little of it, and life on the planet will die; too much and it acts like a blanket, trapping emitted infrared waves that are trying to escape the earth's atmosphere and warming the planet. About five hundred million years ago, before life existed on the earth's surface, plants evolved from water to land, leading to increased oxygen and decreased temperatures and CO_2 levels.[4] Plants absorb carbon dioxide and release oxygen, forming a symbiotic relationship between plants, the sun, water, inorganic nutrients (in the soil), and, eventually, animals. As these plants and algae died, they were buried under the ground and the sea for millions of years and metamorphosized through heat and pressure to become concentrated hydrocarbons. These organic compounds are highly concentrated energy and thus very useful when combusted, creating carbon dioxide, water, and heat. Heat is used for energy, and carbon dioxide is a by-product of the process. These fossil fuels, in the form of crude oil, natural gas, and coal, are energy sources used similarly to the way our bodies use the food we eat to survive. Just as our bodies burn calories, internal combustion engines that burn fossil fuels have an equivalent level of efficiency, around 20 to 25 percent, to generate power. This means that humans and our machines are inefficient in turning the energy inputs into a work output. And since nearly all the food grown on the planet uses fossil fuels in the production and distribution supply chain, our lives are intertwined in the carbon cycle that has led to climate change. Despite the transformational innovation and economic development made possible by fossil fuels, its efficiency rate makes it a primitive form of energy, given what we know today. Humans burn approximately 9.4 gigatons of fossil fuels, and the natural carbon sink absorbs 7 gigatons via the forests (3.0), oceans (2.5), and others, but we're losing 1.6 gigatons through deforestation. In total, approximately 50 percent of the carbon emitted ends in the atmosphere.[5] Therefore, while scientists and policymakers have spent a great deal of time and effort on new technologies and sources for energy, it shouldn't be lost on us that improvements in energy efficiency should be the starting point for solving the climate challenge. Running our built environment from *burning stuff* seems rather crude in the 21st century. Fossil fuels have led to supply chains and innovation, but can't we find a better method in the 21st century?

Here on earth, we have nearly unlimited renewable energy from solar, wind, and other sources, but we haven't perfected ways to harness it and use it more efficiently. Humans and animals have used renewable energy to harness a tiny percentage of energy feedstocks from plants and animals; we eat, burn, and use organic matter in a limited manner, and have done so for thousands of years. For energy, we used to get heat from the sun, fire from wood, and wind to power ships to sail across the ocean, none of which have sufficient energy density to power a modern-day economy. Humans can only do so much with our muscles, as is limited through the use of animals such as horses, camels, and other beasts of burden that were used for transportation and production. The renewable energy of the sun, wind, tides, and geothermal has always been plentiful, providing more of what we need, but we haven't been able to convert it as easily has it has been to burn fossil fuel. At the beginning of the industrial age, we began to use fossil fuels by burning coal to create heat and steam to power rudimentary engines. Later, in the 20th century, we refined oil to power automobiles and factories. Finally, after World War II, the modern energy system had numerous options, including coal, oil, natural gas, and nuclear fission, at our disposal. As a result of scientific and supply chain innovations, fossil fuels and nuclear energy have been used to create a built environment beyond what anyone could have imagined even a century ago. Despite these innovations, we must acknowledge that this crudely implemented scientific process spills dangerously into the planet's ecosystem.

Fossil fuel is effective because it is highly combustible, which also has an environmental downside. Its real benefit as a fuel is its energy density, the amount of power achieved from the source at a certain level of measurement. The measurement is called a joule, defined as the amount of energy of one newton over the space of one meter; a newton is a measurement of the force necessary to move one kilogram-meter per second. The definition of a joule might be a bit confusing, but it is useful in comparing the energy capability of different sources. Humans have an energy density capacity of 1,000 joules per square meter, while oil has 35–45 gigajoules, 4.5 million times more powerful.[6]

Renewable energy sources like the sun and the wind are much less concentrated than even human/animal power in their natural form. If this surprises you, experiment with how much wind it takes to blow over a cup versus knocking it over with a swipe of just one of your fingers. Or stand still in the sun on a cold day and see how long it takes to become warm versus running in the same weather. The sun and the wind can be game-changing potential energy sources, not because they possess greater energy density than fossil fuels, or even humans, but due to their sheer mass, once we learn how to concentrate and harness them from an energy density and scalability standpoint. Unlike fossil fuels, which may last only a century or more and destroy the environment, near-infinite renewable energy can last millions of years. But we need to perfect the science, supply chains, and human systems to make it viable.

The importance of fossil fuels today is unmistakable. In the United States, almost 80 percent of energy is generated by fossil fuels (petroleum, 36.1 percent; natural gas, 32.2 percent; coal, 10.8 percent), with the remainder created by the growing use of renewable (12.5 percent) and nuclear energy (8.4 percent).[7] Consumption is primarily within the industry (32.4 percent) and transportation (27.7 percent) sectors, and the United States is responsible for 17 percent of the world's energy consumption despite comprising less than 5 percent of the world's population.[8] Energy consumption is growing in the United States and worldwide, with overall energy demand outpacing increases in renewable energy resources and improvements in utilization and efficiency.

Despite the media narrative around renewable energy, the US Department of Energy predicts that in 2050 75 percent of the energy consumed in the United States will come from fossil fuels, a puzzling prediction given all the climate promises made by government and industry to become carbon neutral by the same year.[9] This forecast of a slow adoption of renewable energy may have something to do with the scale of America's known fossil fuel reserves, estimated to last another century at today's production rates.[10] I believe this dismal forecast is due to a lack of understanding of the challenges with renewable energy being not its source but rather the generation and distribution supply chain. From one perspective, America has an infinite amount of solar and wind, and there are technologies in place to make renewable energy dominant. On the other hand, nations like the United States are sitting on vast fossil fuel reserves, an energy source that can be monetized, in contrast to renewables. There is an efficient generation and distribution system for fossil fuels, but not for renewable energy. This is the mixed-message storyline of our so-called energy transition, a process that is simpler from a technical than a human systems standpoint.

Energy Calculus

For thousands of years, global energy consumption was left unchanged in burning biomass (e.g., wood) and under 20,000 terawatts a year, or less than 0.13 percent of today's consumption. A terawatt is equal to one trillion watts, and a watt is a unit of power equal to one joule per second. Around World War II, the use of coal doubled, and shortly after that, oil took off, with consumption increasing fivefold from pre-industrial days.[11] Then came natural gas, which completed the range of fossil fuels. In the 1950s, oil economists began to predict "peak oil," when global oil production is at its maximum and demand begins to exceed supply. According to a British Petroleum 2021 report, the worldwide energy use of fossil fuels increased 12.3 percent over the previous ten years, with gains across the board for all types of energy, including the highest growth rates in solar (1,393 percent) and wind (301 percent).[12] However, a deeper dive into the numbers reveals that solar and wind together account for only 32 percent of the volume increase over ten years, with fossil fuels accounting for 60 percent. So for all the good news in the media relating to renewable energy growth, and the earlier concerns relating to the so-called peak oil, the reason energy economists aren't more optimistic about the future isn't a lack of change in sustainable energy but rather the outpacing of total consumption, with fossil fuels consumption driving its growth.

Looking at these data, it becomes clear why the 2015 Paris Agreement is such a failure, with 80 percent of the century target of 1.5° Celsius already exceeded. Other than the pandemic year of 2020, energy use driven by fossil fuels has grown every year since the Paris Conference. The developing world aspires to live the lifestyle of the United States, a nation that uses 17 percent of the world's energy while comprising less than 5 percent of the population. Additionally, considering that carbon emissions are projected to linger for three hundred to one thousand years in the atmosphere (nobody knows for sure), the fossil fuel costs and benefits calculus must begin at their origin in the mid-19th century, not today. The fastest rise in energy use is in the Global South, particularly Asia, which now accounts for more than 45 percent of worldwide energy with a lower per capita and historical contribution. Most conspicuous may be sub-Saharan Africa, which is projected to grow to 20 percent of the world's population by 2050 and is less than 12 percent of the total energy use in the world today.[13] The average energy consumption per capita in sub-Saharan Africa is 185 kilowatt-hours a year, compared to 6,500 in Europe and 12,700 in America.[14] For Africa to develop in a manner that reduces widespread poverty, its per capita energy use will need to grow exponentially per capita and in population growth. Many nations worldwide have falling population demographics and growing energy use, while some nations, mainly in Africa, have both growing energy use and growing populations. If the world's energy strategy doesn't change soon, this will be even more catastrophic to the environment.

The average American consumes more than four times the energy of the average global citizen, and I'm betting that if you're an American reading this book, you're above this average. Many of those in poverty are disproportionately impacted by the costs of climate change, such as floods, droughts, and other events, but do not equally benefit from the advances that are causing it. To make this outdated version of energy math work, there needs to be some combination of an improvement in energy efficiency and utilization from the inputs, a reduction in energy usage in the developed world, and a moratorium on further growth in the developing world, none of which are sufficiently happening. Despite the public, high-level proclamations made by government leaders and multinational corporations about attacking climate change, in viewing the details, there are few incentives for these public and private institutions to actually make it happen. Despite the claims, the math doesn't work in the current model, and if we are really serious about achieving environmental stability, we must come to these realizations.

The Failed Strategies of Futures Past

For global temperatures to stay within the 1.5° Celsius target—a goal already unlikely to be met—a significant amount of existing fossil fuel deposits must remain in the ground: 89 percent of coal, 58 percent of oil, and 59 percent of gas.[15] How can we make this happen when so many nations and regions—including the United States, Canada, Russia, China, India, Australia, Nigeria, Venezuela, and the Middle East—would lose significant public revenues by keeping these reserves in the ground? The geopolitics of fossil fuels is a prisoner's dilemma scenario, where each nation must either trust that all others will follow suit or cheat while knowing that others will also do so.

Even though the potential for renewable energy will achieve demand capacity much sooner than any of the public estimates, the pumping of the brakes on its usage stems from the apparent dilemma of valuable assets in the ground that, in a capitalist system, cannot remain in the ground. The authors of a 2022 paper in the open-access journal *Nature Climate Change* found the net present value of lost fossil fuel revenues and profits at $1.4 trillion and $400 billion in the United States, respectively.[16] The estimated $11 to $14 trillion in fossil fuel assets that might be stranded to address climate change could become worthless by 2036, leading to a massive economic crisis much greater than the 2008 collapse.[17] The question of fossil fuel assets in the ground is an issue nobody wishes to discuss, but it must be addressed if the world is serious about addressing climate change. Saudi Aramco, the world's largest oil producer, which produces over nine million barrels of oil daily, invested between $40 and $50 billion in capital expenditures in 2022, not exactly a sign that they plan to turn off the spigots.[18] Never in the history of the market system has such a trade-off for the public good been made at such a scale. If large public and private institutions fail to address this question at their grand conferences, they aren't taking the problem as seriously as necessary.

Grassroots efforts are underway to track carbon emissions and pressure policymakers and companies to keep hydrocarbons in the ground. For example, the Global Registry of Fossil Fuels, which contains data from more than fifty thousand fossil fuel fields in eighty-nine countries, was launched in September 2022. This registry aims to provide data transparency relating to carbon emissions and energy use. A greater understanding of the environmental impact of each of the more than fifty thousand feedstock locations worldwide is a move in the right direction. Still, these grassroots efforts are meaningless without addressing the fundamental question of the large institutions, as noted above. Despite the marketing efforts undertaken by giant fossil fuel companies and oil-exporting nations, there is a financial disincentive to abandon these fossil fuel reserves. According to a 2019 study conducted by the Climate Accountability Institute, the top twenty multinational and state-owned oil companies have contributed 35 percent of energy-related emissions since 1965.[19] These combined companies have market capitalization value in the trillions of dollars, with Saudi Aramco valued at close to $2 trillion. Despite their public support of strategies to transition to green energy, the economics of shareholder capitalism suggest otherwise. We need a more practical understanding of the problem.

At the COP27 Conference in Sharm El-Sheikh, Egypt, there were registrations for 636 lobbyists in the fossil fuel industry, including chief executives of BP and Total and the incoming CEO of Shell.[20] The solution sold by the fossil fuel industry is carbon sequestration, the process of capturing CO_2 before it can be released into the atmosphere. According to a paper published in *Science*, this solution costs $600 per ton; with emissions of 43.1 billion tons annually, this solution would cost $25.8 trillion per year, more than 25 percent of the world economy.[21] Nevertheless, the US company Occidental Petroleum is moving this initiative forward, building its first carbon capture plant in the United States, with plans for operations worldwide. Carbon capture may be just another public relations plan for keeping fossil fuels as king, but at least it's an effort to move in the right direction.

On the negative side of big oil's strategy are so-called investor-state dispute settlements (ISDSs), which allow shareholders and investors to litigate for compensation for government or private-sector action limiting their investments' revenues and profits. Of the more than 55,000 oil and gas projects worldwide, 10,506 (19 percent) are protected by thirty-four ISDS treaties with financial net present values of up to $340 billion.[22] Most oil-producing nations are subject to these treaties if they take bold steps to address climate change, despite the negative impact these decisions could have on their economies.

Even if the leader of an oil-producing nation mandated that its oil reserves stay in the ground, something akin to political suicide, the nation would face lawsuits from multinational corporations, given the terms of the ISDS treaties. In many cases, these nations from the Global South face acute threats from climate change, such as flooding, land degradation in Nigeria, and severe drought in Venezuela. More than 90 percent of the population in Nigeria and Venezuela live on less than $5.50 a day, yet both nations are sitting on large fossil fuel reserves, relying on these commodities for needed revenues to fuel economic growth. Any reduction in revenue to abate climate change must be offset by suitable alternatives, none of which seem likely. How could these nations step away from their most significant source of income as lower- or middle-income nations? Such a well-intended leader would likely be deposed in a matter of days. Despite what is noted in the media, the development of wind and solar technologies is the easy part of the problem to solve. Another quandary to address beyond the market value of the fossil fuels in the ground is related to the subsidization of the oil and gas industry worldwide. These enormous government subsidies represent a significant percentage of the overall worldwide economy, leading to nonmarket forces impacting more rational decision-making. The media and research platforms emphasize that renewable energy is growing, and it is. Renewable energy has become cheaper and more scalable, but it faces unfair nonmarket challenges because of cartel and government subsidies. Capital markets are beginning to lean toward renewable energy infrastructure projects over traditional oil and gas projects, but not as rapidly as required to address the environmental problem. In the third quarter of 2022, large multinational oil companies raked in record multibillion-dollar profits, with ExxonMobil reporting almost $20 billion and Chevron meeting its highest quarter ever.[23] New oil and gas projects are sprouting worldwide, with eighty-seven new projects (including twenty-six in the United States and eleven in Canada) expected to be operational by 2030.[24] The evidence is overwhelming that our human systems—not science, technology, or the supply chain—are the weakest link in solving the problem.

The geopolitics of the fossil fuel industry cannot be ignored. According to current calculations, there are approximately 1.4 trillion barrels of oil in reserves, nearly forty years' worth of consumption at the present worldwide rate.[25] Not only is there sufficient oil in current known resources to last until around 2063, more than a decade past the dreaded 2050 date, but most of these reserves are also profitable to these oil-producing nations at higher ($90 a barrel) and lower prices ($45 a barrel).[26] The storyline is a great paradox: economies in desperate need of growth, the promise of significant oil and gas use, a considerable war undertaken by one of the world's largest fossil fuel producers, record profits of the oil producers, and the devasting impact of high concentrations of CO_2 emissions. Despite the claims made at these global conferences, this tells the real story of what's happening with CO_2 emissions.

Playing Monopoly versus Changing the Game

Let's be clear about what's happening in today's energy markets: the world could be on renewables much sooner than is currently forecasted if not for these challenges within its human systems. One 2021 study found a significant increase of more than $400 billion in direct subsidies worldwide for fossil fuels.[27] According to the International Institute for Sustainable Development, direct production subsidies for fossil fuels averaged $290 billion and consumption subsidies $320 billion.[28] Other studies calculate the subsidies to be much greater, at $5.9 trillion, or 6.8 percent of the world's GDP, in 2020.[29] The vast differences in these numbers demonstrate the explicit subsidies (noted above) that are direct funds to suppliers and the implicit costs that must be paid related to climate change.

Beyond oil and gas, worldwide electricity markets and utilities are also a complex labyrinth of unclear regulations, subsidies, and accounting that are difficult to understand, much less manage. When Thomas Edison invented the light bulb in 1879, he knew it would be meaningless without an electricity generation plant, so he installed one in New York City in 1882.[30] In many nations, the electric power grid is municipal, owned by states or towns. Nearly all of the electric power grids in the United States are privately owned, a natural monopoly that includes more than seventy-three hundred power plants and transformers connected by more than 160,000 miles of high-voltage transmission lines.[31] While most direct government subsidies for power generation are provided to renewable feedstocks, such as solar and wind, these amounts pale in comparison to the total subsidies provided to the oil and gas industry. Yet the math related to these public and monopolistic utilities is challenging to untangle, but it must be if we are to reform and transform it.

Generating electricity is the easy part of the game: the more difficult part is creating an imbalance of electrons so they flow from one location to another. According to a 2023 report from the Energy Transitions Commission, 55 percent of the investment capital is required for energy generation, and the other 45 percent is for distribution, transmission, and storage, accounting for more than a trillion dollars worldwide annually.[32] Generation of renewable energy, such as solar, can happen through a photovoltaic process using a semiconductive material on the solar panel, usually silicon, that enables an electric charge. An inverter turns the charge into electrical energy on the panel. That energy can be used directly to power homes and buildings, or it can be stored in a battery. The challenge is balancing energy generation, transmission, and use; the sun doesn't always shine, while coal and natural gas can be burned on demand. In the past, power was received from one generator or power station; today, the sources are diverse in both availability and scale. The modern-day electric grid isn't sufficient to manage the synchronization required to manage divergent power sources. Innovation toward renewable sources presents challenges to these monopolies, which struggle with stability, as was evident in major crises in recent years. Therefore, the greatest challenge isn't in generating energy from solar panels or wind turbines, but rather the entire generation to distribution process and how the current utility model must be reinvented; if we solve this problem, renewable energy becomes the game changer.

My first exposure to utility companies was in the game Monopoly, where the player has to roll the dice and pay the utility company what it told you to pay. Monopoly utility companies are the other undiscussed topic, as they are incapable of leading us to a clean energy source in our homes. Formalized through government regulation and utility commissions, these organizations often follow different rules. In the United States, power grid utility entities have different accounting rules where the focus is not on selling the power to consumers as much as it is on a cost justification of the infrastructure investments related to generators, substations, transmission lines, and the like. This 20th-century model focuses on centralized energy generation and transmission, and utility rates are based on feedstock costs and depreciation costs on installed infrastructure. As such, utility companies aren't sufficiently incentivized to innovate in renewable feedstocks and become more efficient across the end-to-end supply chain, including having distributed energy resources (DERs) and networks across providers through the grid.[33] In addition, pricing models for generators are archaic, don't make sense, and often reward inefficiency. These outdated rules worldwide must be addressed to unlock the viability of renewable energy to power the grid.

Beyond innovation, today's utility companies cannot meet current state requirements within their monopolies. For example, the 2018 Camp Fire, one of the largest in US history, was driven by the outdated infrastructure of Pacific Gas and Electric, the largest provider in the country. In 2021, Texas faced a major power grid failure that led to 4.5 million homes and businesses going without power for an extended period during cold weather; hundreds of people died. Even the 2023 Maui wildfire disaster is at least partially related to the public utility Hawaiian Electrical Industries, Inc., in managing its dilapidated power line system.

Given the energy volatility resulting from the Russian invasion of Ukraine, the instabilities of the existing grid, and monopolies and subsidies, progress toward sustainability targets of carbon neutrality by 2050 lacks a necessary focus. Innovations and markets work best when there are alternatives. A decentralized power generation model of DERs focused on solar could be a game changer for a 21st-century sustainability strategy. And yet the Florida legislature passed a bill—vetoed by governor Ron DeSantis—that would have authorized utility companies to charge residential solar customers for lost revenue. In the outdated 19th/20th-century model of a centralized power grid, losing customers could lead to a complete grid failure. Moving forward, these roadblocks cannot be ignored.

The monopolies that exist in the Global North may be a roadblock to transformation, but fewer barriers exist in the developing world, where power requirements are growing the fastest. In this book, I have made a case that it is the responsibility of the richest in the world to fund a 21st-century sustainability strategy for the rest of the planet. Most of the developed world, including 70 to 80 percent of the world's population, agrees with this strategy. Still, there is no consensus on who will pay for it. Should China be considered a middle-class or a wealthy nation? How will citizens of the so-called wealthier nations, such as the United States, perceive investment in other nations when there is so much need in their own? Recent proposals have been brought forward, such as US climate envoy John Kerry's proposed carbon offset plan that allows corporations to fund renewable energy projects in developing countries. Unfortunately, none of these discussions have led to much change, as these proposals are more tax incentives and subsidies rather than direct investments.

Given the lack of utility infrastructure, the growing need for energy, and the disproportionate impact of climate change, developing nations have significant incentives to proceed rapidly toward renewable energy. Yet governments have been proceeding slowly; for example, Egypt set a target in 2016 to produce 20 percent of its energy from renewable sources by 2020, but in 2021, that figure was just 12 percent, primarily due to an increase in demand.[34] Vietnam, a growing developing nation that is one of the twenty largest consumers of coal, had a plan to wean itself off coal by the end of this decade but was forced to revise this strategy backward because of a lack of funding promised from the wealthiest nations at the COP27 Conference, stating now only that it will have a meaningful decline in coal use by 2045.[35] Indian prime minister Narendra Modi has pledged to make his country a green powerhouse by increasing solar and wind use threefold by 2030, but the nation will only reach carbon neutrality by 2070.[36]

In Adam Smith's model of capitalism, where resources are focused on the area of greatest need and opportunity, capital funding for electrical grids would naturally migrate to Asia and Africa, especially given the human systems challenges in the West. If the ESG model of investment is working, the capital raised should be focused on the Asian and African nations where the epicenter of the climate change battle is taking place. For example, energy demand for the ten Association of South-East Asian Nations (ASEAN) is projected to increase by a third of the European Union's (EU) current total by 2050. India alone will exceed the EU's energy requirements by 2040.[37] Meanwhile, large Asian nations such as China, South Korea, and Japan are funding coal-powered plants in these developing nations to meet their energy requirements. Today, the world's energy strategy is moving in the wrong direction, but that could change quickly and effectively with the proper design, which I will discuss in the next section.

Solution: Community-Based Energy Systems

To summarize, untangling today's energy markets and infrastructure from the failures of current institutions would take too much time and is unlikely to be successful. The more practical strategy is to provide legitimate alternatives with a greater emphasis on distributed systems in the developed world, and a focus on areas with fewer impediments to change, such as less developed nations in Africa. The good news is that if we can fix these significant challenges within our human systems, there is a potential for renewable energy goals to be met well before 2050, given advancements in technology. Even though energy subsidies are relics from the 20th century and discourage disruptive innovation, energy innovation can happen soon. Governments that are serious about transitioning to renewable energy should address the impact of their energy subsidies, accounting rules, and regulations. One study showed that fossil fuel subsidies are $5.8 trillion a year, with more than half being from three nations: China ($2.2 trillion), the United States ($646 billion), and Russia ($522 billion).[38] Note that these subsidies include unaccounted environmental costs; if the focus is just on monetary subsidies, that figure is approximately $555 billion, or $584 billion when including public financing of producers.[39] It's challenging at best and impossible at worst to compare and contrast subsidies for this and other industries as a means of understanding their direct impact on the economy and environment. However, providing direct or indirect subsidies for fossil fuel production and consumption is contradictory to these climate pledges that are made annually; governments and corporations need their actions to speak louder than their words.

The starting point for this strategy is to not even get caught up in trying to fight the authority of the global energy sector. The influence of these large institutions of governance is pervasive in the design, distribution, and use of energy in our modern societies. To demonstrate the math of this model, I will use the average US family as the focus for the design, which makes sense because the United States is disproportionate in both consumption per capita and total percentage worldwide, so a solution that achieves success in this country can be applied to any other nation. According to the US Energy Information Administration, the average American family consumes 11,000 kilowatt-hours per year.[40] Based on various studies, the energy required to run an electric car (to wean off fossil fuels) is 4,000 to 5,000 kilowatt-hours per year; for this analysis, I will assume that the average family has 1.5 cars and the energy requirement is 4,500, or an annual total of 6,750 kilowatt-hours a year, making the average US family's yearly total energy requirement 17,750 kilowatt-hours a year.

Powering a house and vehicles with solar energy requires a 20-watt system to provide 20,000 kilowatt-hours a year. In late 2022, the cost of such a system was approximately $55,000, meaning that in comparison to getting energy off the grid and gas at the pump, it would take more than sixteen years to pay back, without accounting for a time value of money, which is infeasible. It also does not consider federal, state, or local tax credits allowing buybacks of excess power from the consumer. Based on this high-level analysis, including the cost of capital for the homeowner to have a solar energy system and some tax credits, the cost per kilowatt-hour is much higher than the cost of using the electrical grid and buying gas at the pump. It can be complicated to make an apples-to-apples comparison, and I haven't seen any studies that have done so without an understanding that, at present, running a residence power off solar is cost-effective. The problem with solar is that the upfront cost to shift from old to new technology, which is beyond the ability of the average American to pay, is only partially offset by federal and state tax credits. According to the Solar Energy Industries Association, 130.9 gigawatts of installed capacity in the United States is enough to power 23 million American homes, or 16 percent of the total.[41] Progress is happening in the generation of wind and solar power, but insufficient progress is being made in developing a 21st-century decentralized-centralized smart grid for renewable energy. Investing in this new model must happen, despite the special interests from the monopolistic utility interests.

As much as politicians speak of a future of renewable energy and carbon neutrality in impressive, progressive terms, there isn't much incentive for even the most prosperous Americans to move in this direction. According to the International Renewable Energy Agency (IRENA), two-thirds of all wind and solar projects that came online in 2020 were cheaper than new fossil fuel projects for the same period—but not necessarily to the end user.[42] This statistic is somewhat misleading because it doesn't compare the efficiency of these projects to existing fossil fuel investments that have been fully or partially depreciated or consider other factors in energy, such as distribution and storage. Moreover, renewable energy *should* be cheaper than fossil fuels in the long term; a study in the peer-reviewed journal *Joule* calculated $12 trillion in global savings by 2050 by switching to renewable energy.[43] The problem is that these savings calculations include the implicit costs of externalities that are absent in corporate financial statements, either being paid by public governments today or to be borne by those into the future. From an environmental, technological capability, and even a financial perspective, the results are clear and obvious in an apples-to-apples comparison to these monopolies.

The alternative to this centralized 20th-century power generation and distribution model is a decentralized 21st-century model that makes it viable for residences and businesses to generate power from the sun or wind. This strategy is akin to separating from the public utility telephone company when we cut the cord and went wireless. The government can equip or even incentivize homes and businesses with sufficiently sized solar panels and battery storage to become primarily self-sufficient if it decides to do so. Improvements need to be made in the efficiency and storage capacity of solar panels to manage them. Both solar power generation and battery energy storage are sufficient to run the average American house, but the upfront cost of the equipment isn't cost-effective, so improvements in technology and government policy should be encouraged and funded. Today's solar panels on homes and buildings are approximately 15 to 22 percent efficient, but under standard terrestrial conditions, the maximum potential efficiency of a panel is 33.7 percent per the Shockley-Queisser limit. This means there is potential for technology for power generation to improve by a third. As for battery storage, advances in battery efficiency and monitoring will both lower the upfront cost of the equipment and improve utilization of power produced, which is an issue currently. A decentralized model of renewable power generation, tethered to a networked, large-scale centralized model revised for the 21st century, would be the game changer that would break our dependency on fossil fuels. Presently, there is growing progress in distributed energy. The Chinese company Huawei released a solution called FusionSolar. This smart-home system combines solar panel generation, energy storage, and "grid-forming" software to stabilize energy and potentially connect to a primary grid and others in a network.[44] Imagine the possibilities of a networked system that connected decentralized, distributed energy generation at one house to other homes and the larger electrical grid, managed through new technologies such as artificial intelligence to manage the overall flow and storage. The good news is that much of the available technology is in place to transition to this model and t is beginning to take shape in some parts of the world, such as Europe.

In the 20th century, a centralized power grid made sense; in the 21st century, given climate goals and technological innovation, a decentralized hub-and-spoke model should be the strategy. Rather than power being generated and distributed from a central location, it should be accumulated and managed across a network. Consider this power grid like the internet, a backbone for energy flow across a region, nation, or planet. The power grid should also be the regulator of massive solar farms that are networked worldwide. Today, the five largest solar farms are in India (#1 and 3), China (#2 and 5), and Egypt (#4), each powering millions of households.[45] Now imagine if every nation in the world had massive solar farms tethered to new, smart power grid networks that enabled these massive additional centralized sources of electricity for larger purposes, such as factories, and fed the balance of the surplus back and forth, with millions of decentralized clusters through residences and businesses. In other parts of the world, such as the North Sea, there is enough wind power to provide energy to most of Europe, including the capacity to create green hydrogen. In the Sahara Desert alone, there is enough space to install solar panels that could power the world. In this scenario of solar and wind, why would we wait until 2050?

According to a 2019 study conducted by Stanford University researchers, there would be a $73 trillion upfront cost to convert to 100 percent renewable energy by 2050, a figure sure to scare the world away from this initiative.[46] Given the existing installed power, the authors of the study created an overly complex and expensive centralized-focused model, yet costs continue to fall precipitously on a smaller-scale community market. Still, for the sake of argument, if this is true, starting in 2026, it would have a straight-line depreciation cost of $2.8 trillion a year, similar to the OECD-estimated cost of the Russian-Ukraine conflict by the end of 2023.[47] With a decentralized-centralized, community-based model, the cost should be significantly less, perhaps by as much as 50 percent. With the global GDP at $103 trillion, and assuming a 50 percent lower capital investment through this community-based model, a five-year investment of $36.5 trillion could cost $7.3 trillion a year based on current technology; with improvements, these costs will fall dramatically. However, instead of looking to cost-justify these sorts of projects on a massive global-scale effort, which is always conceptual and not reality, the business case can be justified on the local, community scale. Simply by eliminating the up-front capital costs for a community-based, decentralized power grid network could happen one brick at a time, like how the internet developed, versus a large, centralized, institutional project that will never happen. As you know, today's internet wasn't built in a day!

The point is that the funding of a renewable energy grid of solar and wind is possible, and capital can be raised to make it happen sooner if the will exists for it. Since it's doubtful that large public institutions would imagine such a macroeconomic scenario, it's up to these community-based initiatives to make it happen through micro- and social entrepreneurship funding. Just as Adam Smith saw the mercantilists of the 18th century as dangerous to economic growth for the public good, we see the same today related to explicit and implicit subsidies. Even without government subsidies for renewables, this is a game-changing model that could reverse the current trajectory of fossil fuels in use as long as 2070, an unsustainable energy strategy. Let capitalism work!

The achievement of this first solution will not happen as a global initiative, and the rationale behind this is already clearly documented in this book. Funding will also need to be nontraditional in source and outflow. We must allow Smith's model of capitalism to work, where resources flow to others with the most significant opportunity, in this case, those areas in the Global South facing the greatest need. From a cost-benefit perspective, the investment would be more productive in the developing world for two reasons. First, given a lack of infrastructure and investment from the oil and gas companies, there would be lower exit strategy costs; and second, these nations are some of the worst and fastest-growing polluters related to fossil fuels such as coal. We shouldn't expect and cannot wait for the most prominent energy-producing nations in the world—such as the Gulf states, Russia, and even Canada and the United States—to be the first adopters. Investment in these community-based energy systems in Africa would be cost beneficial and easy to adopt and could change lives, improve geopolitical security, and move the needle in the right direction for the environment.

The Future Is Bright?

Much has been said of politicians who have promoted an "all of the above" energy strategy, which appears to have first been noted by the Obama White House in 2014. This strategy sounds logical, but it is more of a political slogan than an actual strategy. For a valid renewable approach to be implemented, there must be a focus to scale; trying to take on all options is a lack of commitment to the ones most impactful in making a difference, a jack-of-all-trades and master of none. In my community-based system, the primary feedstock is sunlight and the secondary is wind, given the ubiquity and scale of each. Hydrogen, which can be produced from renewable energy sources, could be a viable portable energy source to replace gas. Others, such as tidal, hydroelectric, thermal, and biofuels, are renewable but not scalable as a focus area for a 2050 strategy. Everywhere in the world has sunlight and wind, and plenty of it; the focus must be on how to gain greater efficiency. Focusing on too many options prevents scalable solutions from happening sooner.

Some may question why I don't include biofuels. Much of biofuel is feedstock that would be better used to feed people rather than to burn for energy. In many cases, biofuels are used not because they are the most efficient but because of government-provided commodity subsidies. The focus should be on supporting farmers, encouraging innovation, and raising efficiencies and global supply chains to make agriculture profitable for farmers to export foodstuffs to other nations in need rather than using food for energy. (I will discuss more relating to this in the next chapter.) Second, biofuels create energy through burning, meaning their use creates CO_2 emissions; although their feedstock sources are renewable, they still have a negative impact on climate. Third, biofuels also require other commodities, notably water, which is a critical resource to manage effectively. Biofuels can play a complementary role, but we should let the most significant and sustainable feedstocks of sunlight and wind be our primary sources. A better alternative option is green hydrogen, which has no emissions, so long as it's made through renewable energy (solar and wind), and especially for its use for portable energy in applications such as transportation.

Alongside wind and solar should be nuclear to reduce our use of fossil fuels. Despite the negative implications associated with atomic energy because of the large-scale, albeit infrequent, disasters that happened at Chernobyl in Ukraine in 1986 and Fukushima, Japan, in 2011, in terms of human health, it is six hundred to one thousand times safer than all fossil fuels and safer than all renewable energy sources other than solar. Furthermore, nuclear energy is more sustainable regarding greenhouse gas emissions than all renewable energy sources.[48] After the Fukushima meltdown in 2011, then-chancellor Angela Merkel and the German Parliament decreed that all of Germany's nuclear power plants would be shut down by 2022; chancellor Olaf Scholz ordered the final three to keep operating after the Russian invasion of Ukraine. With Germany receiving 35 percent of its gas from Russia in 2022, compared 55 percent the year before, it is struggling to manage its economy in light of these actions.[49] According to the Germans, nuclear energy isn't sustainable because of the challenge of securing spent radioactive waste, despite the effectiveness of more than thirty nations in doing so today; other European countries, including Belgium, Switzerland, and Spain, have planned phase-outs.

Nuclear power is falling as a percentage of total electricity worldwide, to approximately 10 percent, down from 17 percent, with its largest usage nations being the United States (30 percent), China (15 percent), and France (14 percent).[50] However, fifty small modular reactor (SMR) projects are being planned worldwide as a next-generation project in the 21st century.[51] While these SMR plants won't have the same economies of scale as the larger 20th-century power plants, they require less capital expenditure and might be safer to manage. An innovation built off the model of the SMRs is the company Oklo, a Silicon Valley start-up with a plan to develop mini-reactors able to use the waste from conventional nuclear power plants.[52] If successful, this will help address the primary concern of operating atomic power: what to do with radioactive waste. Other companies, such as Curio and NDB, are working to develop fuel from nuclear waste for advanced mini-reactors and batteries intended for space that could last twenty-eight thousand years.[53] These innovations in atomic energy make it a solid option in a 21st-century sustainability plan. Beyond nuclear fission, ongoing research is making fantastic breakthroughs in nuclear fusion, but while the potential for this as a near-infinite energy source is exciting, it isn't a strategy that should be a focus for meeting 2050 goals.

The last topic relative to a 21st-century sustainable energy strategy is something that isn't discussed often: the need to reduce energy waste. The United States produces 100 quadrillion British thermal units (BTUs) of energy annually, two-thirds of which is not converted into work; instead, it is just wasted heat.[54] Most of this waste stems from burning fossil fuels, such as gasoline in a car, which only achieves approximately 25 percent energy efficiency. In comparison, an electrified vehicle requires only a little more than a quarter of the energy, with electric cars being

70 percent efficient.[55] Electrification replacing combustion-based power for vehicles and the entire economy would lead to significant improvements, even before the feedstock switches from fossil fuels to renewable energy. Another waste in our energy system is due to something called "vampire energy": power consumed by electronic devices after they are placed on standby or turned off but still plugged into an outlet. According to studies, vampire energy consumption can increase usage by 30 to 40 percent.[56] This waste is expensive and bad for the planet; according to a study conducted by Earthday.org, 100 billion kilowatt-hours of energy is wasted, producing 80 million tons of CO_2, equal to the emissions of 15 million cars.[57] The use of smart plugs and devices must focus on reducing this energy waste. Why not emphasize a greater utilization of existing energy generation through smart technologies? We need to become more efficient in energy use as we transition to renewables.

Finally, we need disruptive innovations for our energy future to be bright. The three that I believe will help us achieve our goals are quantum computing, artificial intelligence, and battery innovation. Quantum computing provides breakthrough opportunities in computational power, imaging, sensing, meteorology, and communications.[58] One main area of opportunity related to the community-based-system solution is the use of quantum computing and artificial intelligence to manage a complex array of nodes within a system of residences, factories, high-rise office buildings, and solar farms across localities, states, regions, and nations to best produce and distribute power based on supply and demand. In contrast to conventional computers, which process based on zeros and ones, quantum computers can process information using zeros, ones, and everything in between. Managing a complex energy system of decentralized centralization without advanced computing will be impossible.

The most significant game changer we need toward a 21st-century sustainability strategy is not energy production or distribution but storage. And how we define the terms "battery" and "storage" needs to be redefined. Today, when you hear the word "battery," you might think of a traditional alkaline battery or even the rechargeable lithium-ion battery in your smartphone, computer, or electric vehicle. However, the future of energy will likely use various solutions, such as solid-state batteries, chemical energy carriers, thermal storage, and gravity storage.[59] One example of gravity storage used with renewable energy is technology from Energy Vault, which builds a rising crane tower that uses excess renewable energy to lift 35-ton concrete blocks and release them when the power is needed.[60]

If you're a pessimist, you have plenty to worry about given how tethered our human systems and economies are to fossil fuels. And regardless of the promises made by the oil and gas companies regarding innovations such as carbon sequestration, we cannot achieve carbon neutrality by 2050 without keeping these hydrocarbons in the ground. It's just a math equation. On the other hand, if you're an optimist, you have much to be happy about—if the goal isn't to take on the large public and private institutions that control today's energy strategy. Instead, we need to rely on alternatives in science, technology, supply chain, and finance to bring forth this new strategy. The capabilities are there much sooner than we can imagine if we have the human systems to make it happen.

Chapter 7

The Future of Food

The first restaurant in the world is said to have opened in Paris, France, in 1765. At the front of this establishment read the sign (translated from French), "Come to me, those whose stomachs ache, and I will restore you."[1] Similar establishments opened across Paris in the years before the French Revolution, intended to restore health to their customers (thus the name "restaurant"), but only the rich could afford the high prices. In those days, millions of French citizens were starving, and their stomachs ached, leading to their revolution in 1789. The French Revolution reverberated across the established societies of Europe, which viewed their wealth and privilege as their right of inheritance. One of those individuals was the 18th-century English economist Thomas Malthus, a cleric-noble who raised concerns about these uprisings in his book *An Essay on the Principle of Population*, published in 1798. Malthus is well known to modern economics students for his theory that food production could grow only arithmetically, while the general population could grow geometrically. Malthus viewed growth in food production as futile because the increasing population would eat right through it, leading only to greater misery and poverty. Malthus believed the poor caused their own suffering and impacted the prosperity of the rest of the population. He believed the surplus population was a scourge to the food system and, as a result, impacted everyone else, including his fellow nobles.

Food has always been a central theme in enabling stability or driving revolution in societies, such as the triggering point for the French Revolution. According to historian Sylvia Neely, a French worker spent half of his salary on bread until there was crop failure in 1788 and 1789 (the year of the French Revolution), when the cost rose to nearly 90 percent of wages.[2] The English revolutionary Thomas Paine joined in to support the revolt in France, writing the *Rights of Man* in 1791 as an appeal for the human rights of the masses and compassion for the poor. It was an interesting era that pitted inheritance, mercantilism, and tradition against the universal rights of humankind. Conditions are better today, yet approximately 9 percent of the world's population faces extreme poverty with food issues, nearly 350 million face acute food insecurity, and 50 million are on the brink of famine.[3] Today, the world grows enough food directly to feed 150 percent of the world's population, but 17 percent of it is wasted while the remaining balance is used for animal feedstock and biofuels, among other nonfood products. Malthus couldn't have predicted the modern supply chain systems that could make possible food production that exceeds population growth; on the contrary, he relegated the responsibility for the problem to the divinity of God through pandemics, famines, and other natural and human-made events that acted to balance the population. In contrast, Paine viewed it as a religious duty for the fortunate to show justice and mercy to all people, especially those in need. Today, food remains a major crisis in the world, as more than 800 million do not get enough to eat, of which 60 percent live in countries in conflict.[4]

The topic of food and population has social anthropological roots reaching back to a time when the size of an organization such as a clan, tribe, or community was determined by the size of its food capacity, and wars were fought over resources. Despite the irrational nature of markets to balance supply and demand and the economy and the environment, some environmentalists infer that the problem is the victims' rather than the perpetrators'. In his 1968 book *The Population Bomb*, Paul Ehrlich predicted worldwide famine due to overpopulation, a neo-Malthusian theory that inadvertently points the finger at the Global South, whose inhabitants are powerless to control their fate. Inferences from modern-day environmentalists are more subtle but like those of Malthus, who proposed that the poor marry later and have fewer children. The modern food production system has advanced, but the weaknesses of our human systems still focus on the symptoms rather than the root causes.

The cruelty and ignorance of blaming the poor for their poverty was not lost on Charles Dickens in 1843 when he wrote his novella *A Christmas Carol*, with Ebenezer Scrooge noting that it was "better the poor should starve and thus decrease the surplus population." Perhaps a modern-day embodiment of Scrooge is the late Prince Philip of the United Kingdom; in a 2008 interview, he blamed higher food prices not on the waste and profit systems that produce 150 percent of what is required and result in 30 to 50 percent misplacement and waste, but on too many people.[5] Twenty years before this, in another interview, the prince noted that if reincarnation exists, he wished to come back as a deadly virus to help reduce overpopulation.[6] Today's neo-Malthusians are a strange mix of aristocrats, conservatives, environmentalists, and even conspiracy theorists, aligning on this central theme of food, population, and the environment. Upon his death in 2021, Prince Philip was championed as an environmentalist by the World Wildlife Fund, an organization he led as president for five years.[7] At her last climate change conference before her death, COP26, Queen Elizabeth II praised her husband for lighting the flame of environmentalism that lives on through her son King Charles and her grandson Prince William.[8] This is not to say that most environmentalists are callous enough to blame the poor for their fate and the environmental crisis; still, the correlation between hunger, poverty, human conflict, and the environment shouldn't be lost on us.

Environmentalists have made direct or indirect statements relating to a healthier and fairer planet by reducing the number of inhabitants. Along with *The Population Bomb*, books by Garrett Hardin (*The Tragedy of the Commons*, 1968) and the Club of Rome (*The End of Growth*, 1972) discuss human population growth as a threat to the planet without regard for its root causes.[9] A theme of the economy and overpopulation as the leading cause for climate change is not only unfounded in the data but actually in direct contrast to what the data are showing: greater economic activity to bring the poor out of poverty will lead to lower population levels. Dickens hinted at the root cause in *A Christmas Carol* when the Ghost of Christmas Present introduces Scrooge to two dirty, poor children he believes belong to the Ghost. Instead, these two orphans belong to humanity; the boy is called "Ignorance" and the girl is "Want," with a particular focus on the boy. This passage is Dickens's perspective that ignorance within our human systems is the root cause, and want (poverty) is the symptom. Rather than focusing on curing the environment by reducing the population, the focus should be on reducing the poverty that leads to problems such as hunger, which are bad for the environment. The data I will discuss in this chapter provide the clear picture.

The Global Food System

Solving the problem of food is perhaps the most significant challenge relative to climate change and meeting carbon neutrality by 2050. There are certainly challenges to solving the energy problem addressed in the last chapter, but the technologies are in place for suitable alternatives; this is not the case for the food system. Food production and its supply chain system account for at least a third of all global CO_2 emissions, with the carbon footprint rising because of population increases and the per capita consumption of meat. Furthermore, food insecurity is the most significant cause of serious conflict, perpetuating the environmental crisis through instability. Mostly, when the topic of environmental stability is discussed, it is only related to energy, and not these other challenges; an equal focus must be undertaken for a food and water strategy.

To study this food problem, I used public data from the UN Food and Agriculture Organization (FAO) and the US Government Accountability Office (GAO); the FAO data are from 1961 to 2020, broken down by country, region, world, and other factors.[10] Globally, the data show that population growth has been higher than food production over the last sixty years (82.4 percent versus 55.5 percent), twenty years (49.1 percent versus 35.4 percent), and ten years (26.6 percent versus 17.8 percent), yet food production remains sufficient for 150 percent of the world's population. Ironically, food production is both more efficient and more wasteful as the population grows. Technologies and modifications in effectiveness, yield, and distribution could not have been imagined by any 18th- or even 19th-century economist; from this perspective, Malthus's theory becomes irrelevant. The problem is no longer a Malthusian trap of higher food production leading to a more significant population and thus greater poverty. Instead, higher food production and higher waste through global supply chains is a problem not of science or technology problem but of the human systems and how our global supply chains operate.

In the aggregate, food production has kept pace with a population increasing at the rate of 260 percent between 1960 and 2020. In regions of population decrease, food production and availability have grown; for example, China's population growth was 53.4 percent in the last sixty years, 10.3 percent in the last twenty, and 4.9 percent in the last ten, compared to food production growth of 93.1, 59.3, and 31.9 percent, respectively, during the same periods. The second most populous nation, India, saw a less dramatic but consistent comparison of the population to food growth in sixty years (66.7 percent versus 85.7 percent), twenty years (23.4 percent versus 60.1 percent), and ten years (10.6 percent versus 28.0 percent). In the developed world, most notably the West, population growth rates have slowed to single-digit levels—1.3 percent and 6.6 percent, respectively, in the EU and the United States over the past ten years—while food production has grown more than 10 percent in an already established system. From these data, the relationship between food production growth and the population is the exact opposite of what Malthus hypothesized (and neo-Malthusians perpetuate); population growth is falling in regions where the food system is healthy, which means that societies and the environment can stabilize when food systems are used for economic growth. This doesn't imply that the most populated nations in the world—India and China—do not have food challenges, because they do; instead, it suggests that there is a greater ability for self-reliance alongside a stable global import/export market to feed their populations.

The transformation of the import/export model of supply chains shows an interesting perspective of food systems. In the past sixty years, imports and exports grew by 81 and 81.5 percent, respectively; in twenty years, 46.2 and 47.5 percent, respectively; and in ten years, 39.7 and 40.6 percent, respectively. This tells us that private supply chain systems have increased food production and the fluidity of products beyond national boundaries. For example, if food is grown in one nation but can fetch a higher price elsewhere, there is no sovereignty regarding the flow of goods within a global supply chain. In a market-based system, an importing nation willing to pay a higher price than a country in greater need will be able to purchase the food when demand is greater than supply. A recent example is the fallout of the Russian invasion of Ukraine, which impacted the poorest nations that rely on Ukraine for its food. At the same time, richer countries, such as China, can purchase surplus commodities given tighter supply. The flow of the global food supply chain is market based, which has led to exponential growth in production but also to an increase in waste, inequality, and expansion of food use beyond direct consumption. These reasons justify the imbalance within the food system relative to global supply and demand, economics, and the environment.

A transactional food supply chain also doesn't prioritize what an end user will do with the feedstock, as the goal in a market-based system is to fetch the highest price possible. Because food production is largely privatized, it is sometimes optimal to sell commodities for consumption by livestock, even while hundreds of millions don't have enough to eat. For example, food production for use as animal feed has multiplied as follows: over sixty years, 55.5 percent overall growth and 45.5 percent animal feed growth; over twenty years, 35.5 versus 24.4 percent, respectively; and over ten years, 17.8 versus 26.1 percent, respectively. Before industrial food production, livestock grazed on grass in open fields, but today, livestock are fed agricultural grains within large supply chain systems. Through more efficient supply chains and the use of economies of scale for grain production to feed livestock, more of the world's population can eat meat, leading to a higher carbon footprint for food. In a market-based system, this can lead to higher prices and a movement of agriculture toward those who pay the highest price. Another use of agriculture is nonfood related, such as its use in biofuels; these other uses of food grew faster than food production over sixty years (55.5 versus 93 percent, respectively), twenty years (35.5 versus 78.1 percent, respectively), and ten years (17.8 versus 12.4 percent, respectively). Privatized and subsidized large food supply chain systems have substantially increased food production beyond population growth, worsening the climate change crisis.

In the past, food was grown through small farms and government control, while today's market approach has led to greater production and waste. In the past, subsistence farming led to little to no food waste because the individuals within a community relied on their local supply within a networked trading system. In today's globalized supply chains, food producers are disconnected from their communities. Waste within an industrial scale calculation is a cost of doing business, often enabled through governmental subsidies. Food waste across the global system has grown faster than the pace of production growth in the past sixty years (55.5 percent food growth versus 67.1 percent food waste growth), twenty years (35.5 versus 45.8 percent, respectively), and ten years (17.8 versus 42.7 percent, respectively). These data are consistent with studies from other sources that indicate food waste across the end-to-end supply chain, driven by market factors, is as high as 50 percent of what is grown. It is a failure of human systems for food waste to have a positive correlation with the market growth and optimization that is causing the challenges of poverty and the environmental crisis.

The present food production system has inherent geographical inequities. For example, the developed nations in North America, Europe, and Asia have declining populations and increases in production, import/export, yield, and crop utilization. On the other hand, the BRICS (Brazil, Russia, India, China, and South Africa), other than Brazil and Russia, have declining rates within larger populations, increasing yet insufficient production, and a growing, influential role as importers and exporters in the global markets. The remaining nations, many in the Global South, have a persistent problem of dependency on imports and an inability to contain their growing population levels.

Food inequalities exist across national borders and even within a single country, with a coefficient of variation as low as 0.17 in the United States and Japan and as high as 0.30 in India. Food disparities in regions such as Africa and the Middle East have become flashpoints from a socioeconomic, geopolitical, and environmental perspective that will continue to lead to escalating conflict. Despite the effectiveness of global supply chains in the import/export markets, communities without self-reliance in their food systems are vulnerable to more significant issues, as shown in today's world in regions such as sub-Saharan Africa and southern Asia

Before the COVID-19 pandemic, the issues of population, poverty, and food production were improving in Asia, eastern and southern Europe, and South America, but not in Africa. At best, Africa's growth in food production has kept pace with its population growth over the past sixty years (population growth of 78.4 percent versus food growth of 82.2 percent), twenty years (39.5 versus 51.7 percent, respectively), and ten years (22.5 versus 25.7 percent, respectively). The growth of Africa's dependency on imports is necessary to keep pace with its population growth, with which it has barely kept pace. Embedded in these data is the burgeoning growth of a middle class in fast-growing African nations, leading to greater food inequalities within these nations and across the continent, an example of failures within the present market-based system. From these data, one nation classification that was conspicuously the worst performing is "small island developing nations." In the past sixty years, their populations rose by 62.6 percent while their food production grew by 39.9 percent; over twenty years, it was 31.0 and 3.1 percent, respectively; in the last ten years, it was 23.5 and -21.1 percent, respectively. These data provide clear evidence of the strong correlation between a failure of markets and the impact of climate change on the well-being of these small island nations. A similar case is Africa, the region most impacted by climate change despite having the lowest contribution to greenhouse gas emissions.

More than two hundred years ago, Thomas Malthus theorized that growth in food production was pointless because the population would always grow faster. Current data show this theory to be incorrect; the poor in the Global South aren't the cause of their misery and damage to the environment. Rather than a negative correlation between poverty and the environment, where it is assumed the former is a cause for the latter, the connection between the two has proved to be positive: improving the economic status of the poor, as has happened in China, is the best safeguard possible for the environment. When food production is industrialized and included as part of a community-based system, as I suggest in this chapter, both problems are addressed. Globalized food supply chains have improved scale but not equity or environmental sustainability; the new model will achieve scale differently, more equitably, andsustainably.

Water, Water, Everywhere?

Here's a riddle: how can you put more than a gallon of water in your mouth and have plenty of room to spare? Answer: eat one almond. Different studies have estimated that it takes more than a gallon of water to grow a single almond, with one study calculating that it takes more than three gallons![11] And yet, despite the enormous amounts of water used to grow almonds, two-thirds of the world's population will face some water shortage by 2025.[12] If the planet is under such dire straits relative to water resources, is it feasible to continue growing crops like almonds? Or is this water crisis not a crisis at all? The answer to this question is yes and no; yes, the world is experiencing a water crisis as it is defined today, and no, there is no water shortage if we redefine the term itself. We live on a planet full of water; 71 percent of the earth's surface is the ocean, which holds 96.5 percent of the planet's water. The problem is that ocean water is salinated, meaning we cannot use it for consumption unless it is desalinated, an effective but highly energy-intensive process. Of the nonocean water on the planet—3.5 percent of the planet's total water—0.9 percent is salinated, leaving only 2.5 percent of the planet's water as nonsalinated. When I take you through the rest of the math, you'll learn that if we use our resources properly, this leaves us plenty of water to meet the needs of everyone on the planet.

It's important to note that the planet cannot run out of water because the earth's water cycle is a closed-loop system. Water is not lost or gained but repurposed indefinitely; the water you drink today has been around for hundreds of millions of years in different forms. This means that if there's a drought in Africa or California, it is not a loss of water per se but a challenge of available freshwater resources for that geographic area or a reclassification from fresh to nonpotable water. Of the freshwater on the planet (2.5 percent of total water), 68.7 percent is in glaciers and ice caps, 30.1 percent is groundwater, and only 1.2 percent is surface water. Of that surface water, 69 percent is ground ice/permafrost, nearly 21 percent is in lakes, and the remaining is soil moisture (3.8 percent), swamps and marshes (2.6 percent), rivers (0.49 percent), the atmosphere (3.0 percent), and living things (0.26 percent).[13] What many of us consider today to be "water," based on our limited definition of the term, accounts for 0.007 percent of the earth's total water; when defined as comprehensive sources of available water, including what is trapped in glaciers and snowfields, that percentage rises to 1 percent.[14] These numbers provide evidence that humans are using only a tiny fraction of water available—and even worse than that, we are using it in a horrible, wasteful manner!

From these data, there are three key themes to understand. First, with a world population of eight billion, our conventional definition, use, and reuse of traditional potable water must change. Humans must develop a more comprehensive and efficient strategy to utilize our conventional water sources, mainly groundwater, lakes, and rivers. Second, we must find technologies and processes to bring nontraditional sources, both salinated and nonsalinated, into our water supply chains. And third, none of this will be possible without innovations in how energy and food are developed/grown, utilized, and distributed. Simply put, humanity's greatest challenges with water are related to our industrial supply chains.

Freshwater use has increased eightfold since the beginning of the 20th century, from approximately 500 billion cubic meters to more than 4 trillion cubic meters in 2009.[15] These trends have a strong correlation with industrial food production and its use of water: a disproportionate use of water commenced and continues in the prosperous Global North. Still, usage in the developing BRICS is rapidly increasing, with no end in sight. The most severe problems exist in the poorest nations, such as those in Africa and developing small island nations that face challenging geographical conditions exacerbated by climate change. According to a UN FAO study, less than 17 percent of water withdrawals worldwide are related to the industry. Still, the percentage is growing in developing nations; in Europe and North America, it ranges from 70 percent to more than 90 percent, and in the United States it is 50 percent.[16] As countries develop, there are improvements in water management, but industrial activities in farming, ranching, and other manufacturing sectors lead to a more extensive water consumption footprint. The good news is that modern processes and technologies can improve water quality and availability worldwide, but progress also leads to greater consumption, an issue that needs to be addressed.

By fixing these issues in our industrial systems, including increasing nontraditional sources, not only will humans not run out of water, but additional resources will become available. On the other hand, if water continues to be defined narrowly as the existing system of freshwater that is improperly managed, there will be a growing water crisis. However, with a paradigm shift and understanding that we live on a planet with a closed-loop water cycle system, the only threat we face is a lack of knowledge in solving the problem. In Samuel Taylor Coleridge's 1834 poem *The Rime of the Ancient Mariner*, the main character recalls, "Water, water, every where / Nor any drop to drink." The ancient mariner was trapped in the ocean without options, but today we should no longer consider ourselves captive to the past.

The Problem of Food

Like cars, humans run on energy. Our energy comes from calories in the food we eat, which our bodies burn to generate the energy required to perform essential bodily functions such as blood flow, breathing, thinking, walking, and so forth. In general, a human being requires between 2,000 and 3,000 calories a day; at a minimum, a person can survive on 500 to 800 calories a day for a short period. We require food for both its calories and the nutrients it provides, meaning not all calories are the same. For most of history, humans had to eat whatever was able to be locally gathered or grown, and even today, a significant percentage of the world's population has limited options based on supply. According to the UN FAO, there are substantial disparities in caloric intake from one country to another, with 40 percent of nations over the 3,000-calorie threshold and 12 percent at or below the caloric intake of 2,300 calories defined as the poverty level. The disparity of caloric intake per capita shows a strong correlation worldwide in economics per capita. Furthermore, of the nearly 50 percent of nations with an average caloric intake within the healthy range, there is significant income inequality, meaning a large percentage of the population is also under the threshold.

Given the uneven distribution of the food system and the growth in consumption of processed foods, higher caloric intake is becoming an equally disturbing problem. In the United States, which has an average caloric intake of 3,782, studies have shown an increased correlation between poverty, inequality, and higher caloric intake.[17] Today's industrial food supply chains focus on producing large volumes of processed carbohydrates that are supply chain efficient but not sufficiently healthy. More than 90 percent of crop variety has disappeared, half of the breeds of domestic animals have been lost, and all of the world's seventeen main fishing grounds have been depleted beyond their sustainable limits.[18] Today, 75 percent of the world's food is generated from twelve plants and five animal species, driven primarily by production and distribution efficiency. Approximately half of the people on the planet rely on three crops for their subsistence: rice, wheat, and corn. An increasing percentage of the world's population is dependent on processed food that is cheaper and more scalable to meet the needs of a growing population. Our variety of options has risen, but our consumption quality has not.

Food has come to the center of geopolitical and environmental conflict. The breadbasket of Ukraine and Russia is impacted by war, with Ukraine practically shut down in 2022 compared to its 2021 production of wheat, corn, and sunflower, which comprised 10, 14, and 47 percent of the world's exports, respectively. Some countries—including Russia and Australia, which had record harvests—picked up the slack, but other regions such as China, Europe, and the United States had less-than-stellar yields in 2022 due to climate change. The loss of production was felt most significantly in developing nations, where Ukraine provided food for 400 million people, and parts of the Middle East and Africa. Global food commodity markets have been impacted by the Russian invasion of Ukraine, primarily hurting the poorer nations of the Global South and further perpetuating poverty. Given its food production shortfalls, China has also been reported to hold 69 percent of the world's corn reserves, 60 percent of rice, 51 percent of wheat, and 37 percent of soybeans.[19] As some nations have had less effective crop yields due to climate change, they have used their economic strength to buy more on the open market, which affects other nations that are heavily reliant on imports and philanthropy. Future shocks to the food production system are already happening because of supply shortages of nitrogen, phosphorus, and potash, primarily due to the Russian invasion of Ukraine.

A significant percentage of the food problem is the amount of waste and diversion from human consumption, which will be discussed in the next section. According to a report from the United Kingdom by the World Wildlife Fund and Tesco, about 1.2 billion tons, or 31 percent, of food is thrown away, with 58 percent of this happening on farms alone.[20] The most significant reasons for food loss are government policy toward agriculture and the economics of the supply chain system. Agriculture is a foundational industry critical to nearly every nation's economic success and political stability. All the top fifty-four economies subsidize their farming industries, with a combined total of $700 billion a year in subsidies, or 12 percent of total farm revenue.[21] In these models, farming is a protected industry and the economic viability of the farmer sacrosanct. Farmers sell their produce on the open market, where prices fluctuate. Given the sociopolitical importance of food from both a supply and a demand standpoint, this is a highly subsidized industry worldwide, leading to significant waste. In the private sector, food manufacturers, distributors, and retailers assume high levels of product waste in their profit model, especially in high-consumption nations such as the United States. Consumers in the developed world expect perfect produce and sizable portions in retail and restaurants, with food loss as simply a cost of doing business. Therefore, government policy and modern supply chains have led to a system where food insecurity could be resolved by the same amount of food wasted, approximately 30 percent.[22] As a result of these inefficient and inequitable policies, prosperous nations in the Global North waste more food than they import, and Americans spend less than 15 percent of their earnings on food, compared to 40 to 60 percent in poorer Asian and African nations.[23] (Note that these statistics were developed before the COVID-19 pandemic, and results are likely even more inequitable today.) The data clearly show that the problem with food is not our production techniques or technology but our public and private human systems.

Of Cattle and Cars

To summarize, food production has grown consistently with population growth, with global production sufficient to feed 150 percent of the world's population, yet between 12 and 25 percent of the population doesn't get enough calories.[24] Furthermore, today's market-based system and government policies cause significant food and water waste, estimated at 31 percent of the total—a market failure. And according to the UN FAO, only 48 percent of food grown is directly consumed by humans. Almost 41 percent is fed to animals, mainly for meat production, and the remaining 11 percent is for industrial use, primarily for biofuels. More produce is used to feed cattle and cars than is provided directly to humans, leading to inequality and adverse environmental impact. For those in the Global South, this means they are hit once with the problem of obtaining sufficient and affordable food and then again with the disproportionate impact of climate change.

An imbalanced food pyramid, in this case, can be defined as follows: at the baseline of the pyramid is food production for direct consumption, its original purpose. The next level is food production fed to livestock for meat production. The growth in this second tier is driven by an industrial Global North model of ranching and meat production versus the smaller, traditional smaller farming method of raising grass-fed livestock. And at the top of the pyramid are grains for nonfood industrial output, such as biofuels. According to some estimates, crops used to create biofuels could feed 1.9 billion people, almost 25 percent of the world's population.[25]

The profile of this food pyramid geographically around the world is as follows: at the lowest levels of the pyramid are the developing nations mainly in Africa and developing island nations, which continue to utilize 70 to 90 percent of their crops for direct consumption. The next rung in the pyramid are the BRICS nations, which are moving more toward meat production and consumption; China, for example, is now using less than half of its crops for direct consumption, compared to 75 percent in the 1960s. Finally, European and other developed nations are now producing less than 30 percent for immediate consumption. In the United States, with its high meat production and nearly 50 percent of its crops used for biofuels, that figure is approximately 10 percent. In G20 nations, strategies for the use of crops are driven as much, if not more, by profit and price stabilization, including subsidies, for farmers and ranchers as they are by a focus on feeding the world. Thus, the socioeconomic and environmental problems of food aren't related to our ability to grow a sufficient volume to feed eight billion people as much as the transactional nature of private exchange that makes it profitable to waste 30 percent and to utilize more than half of what is sold for purposes other than direct human consumption. The privatized large-scale global supply chains have led to greater supply but aren't necessarily benefiting the public good through less hunger and environmental sustainability.

Ancillary problems associated with higher levels of meat production and consumption are higher methane and CO_2 emissions; greater use of antibiotics in factory farms, which diminishes their effectiveness in humans; and increases in the potential for animal-to-human viral transmission, which was likely the root cause of the COVID-19 pandemic. With a growing percentage of the world's eight billion inhabitants eating meat, this trajectory is unsustainable under current supply chain practices. Concurrently, overproduction and monoculture are resulting in degraded soils incapable of growing crops without the use of industrial-scale synthetic chemicals that are environmentally harmful and geopolitically contentious. According to the UN Convention to Combat Soil Desertification, soil erosion could reduce crop yields by 10 percent by 2050, leading to $23 trillion in losses.[26] The use of modern agriculture and its market failures of overproduction, waste, and misallocation have degraded a large chunk of the planet from its natural state. The UN Convention to Combat Soil Desertification estimates that 70 percent of the earth has been altered from its natural condition, with Asia, South America, and Africa being the most damaged.

Human systems and our transactional markets focused on price rather than value are the root cause of what's happening to the environment. The problem isn't population growth, as too many have hypothesized, but rather market failures that are not in line with Adam Smith's balancing of development and public benefit. We need a new approach to food supply chains and their markets, and I will discuss these solutions in the sections below.

Solution #1: Human Systems: Networked Community-Based Supply Chains

Food supply chains for the 21st century need to advance beyond simplistic strategies from the 19th century that fail to recognize these crises of starvation and environmental damage. With the right approach, there is more than enough food and water for all the earth's inhabitants. Current strategies drive inequality and planetary damage. Properly distributed and managed, food production can enable stability not just in the aggregate, but within nations, regions, and communities. As noted in the data in this chapter, a scalable and ubiquitous production and consumption model can address the challenges our food system faces. It needs to be market based in the way Smith defined markets: identified in local nodes of community-based supply chains that can serve all communities in a networked glocal system. It needs to be a model where African nations and developing poor island nations have self-reliance and available resources within a global market that optimizes economics and the environment.

This networked model is consistent with my proposal for a 21st-century energy strategy featuring individual solar power generation and storage connected to sizable solar energy farms. In this case, I am proposing efficient, large-scale, sustainable food farms interconnected to community-based food systems in a more sustainable, competitive, and entrepreneurial model than today's highly subsidized institutionalized systems. The current agriculture system is built for economies of scale that are overly dependent on import and export markets and government subsidies rather than free-market measures and local self-reliance. Moreover, in today's food system, the nexus of control is based on global financial transactions rather than the best interests of individuals and communities from a value and resiliency standpoint. Replacing this antiquated 19th-century model with a networked community-based design tied to extensive industrialized farming can create the balance needed for efficiency, localization, resiliency, and sustainability. Community-based systems networked to the more extensive system represent the optimal model to stabilize food supply and demand while stabilizing the environment for sustainability.

At global climate conferences like COP27 in Egypt, significant amounts of money were promised to developing nations through so-called loss and damage funds for the most vulnerable and damaged countries from climate change. Furthermore, ESG investing is intended to be deployed to the most suitable market opportunities to balance investment and the planet. Proposals should be written to fund these localized, community-based food supply chains for energy (discussed in the last chapter), food, and water in desperate places such as sub-Saharan Africa. Self-reliance is a critical value within the human systems in these small islands, in Africa, in inner cities and rural towns; poorer nations and communities have not succeeded by relying on global supply chains. As I will discuss in the following section, investments in these systems should replace government subsidies focused on the existing system. If intergovernmental, national, and environmental groups are serious about improving the economy and the environment as well as reducing poverty, they must support these systems that will achieve local self-reliance. Otherwise, these public policies are just talk.

Solution #2: Big Farming: Reform Water Rights and Farming Subsidies

Given the tight relationship between government subsidies and the agriculture and ranching sectors, it is no surprise that the practice of politicians embedding subsidies and other spending projects into government budgets is pejoratively called "pork barrel politics." Despite spectacular gains in efficiency, production, and sheer volume, today's agribusinesses are closed-market systems focused on market concentration, which discourages innovation through subsidies received from public funding. From one perspective, government subsidies are critical to ensuring an essential industry; on the other hand, they represent a system where the government picks the winners and the losers in a way that crowds out innovation and sustainability. If all developed nations have agriculture subsidies, is this version of the prisoner's dilemma more harmful than good? Economists and politicians shouldn't quote Adam Smith in touting free markets on the one hand while they subsidize an entire industry on the other.

Human systems, not the lack of technology, are the most significant roadblock to a 21st-century sustainability strategy related to food. According to a recent United Nations report, 90 percent of the $540 billion in farming subsidies increase food inequality and harm the planet.[27] Of course, farming and ranching are challenging sectors that include significant risks related to weather and global markets. Providing assurances to ensure stability in food production is essential. When discussed in public, it is presented as help for small family farmers; in reality, a quarter of US farm subsidies went to the top 1 percent of farms, and two-thirds went to the top 10 percent.[28] The argument that these large farms and ranches are responsible for most of the nation's food production is mainly accurate, but other than politics, what benefit are these public funds providing to enable innovation of improvements in production, distribution, and sustainability? Public funds should be justified annually for their costs and benefits. Reallocating these subsidies for activities more beneficial to local economic interests and the environment would be a game changer in setting up the community-based systems discussed in Solution #1. However, this would require courage and reform in the human systems of politics.

In 2021, the EU Parliament approved an initial attempt at subsidy reform that was more significant than anything implemented worldwide in decades. Still, these efforts in Europe are merely a starting point for the transformation that is necessary to liberate the food systems. It is true that the will of the people in a world where approximately 70 percent of the population faces some level of relative poverty will prioritize food prices and availability over climate change policies in the short term, so political leaders and policymakers develop solutions that stabilize food prices, the production process, and the environment, as I am proposing in this chapter. For all the grand speeches at climate change conferences that are politically marketable, undertaking activities inconsistent with these messages is disingenuous.

Hoping for change and technological improvements isn't sufficient; there needs to be a complete reformation of public policies toward food subsidies and water rights. First, the public must be told the truth of what's happening rather than confused by doublespeak and greenwashing policies. Food and energy government subsidies are the proverbial "elephant in the room" not discussed in any of these great global climate crises, including by the environmentalists. Agribusiness subsidies mask the actual cost of food and the inequity of its distribution even before factoring in the environmental cost. Rather than this government policy, more efforts should be undertaken to spark food and water supply chain innovation that is necessary for a 21st-century economy and sustainability strategy. There needs to be an honest discussion of what's happening; otherwise, nothing will change.

Water use is the most significant sustainability challenge for plant-based production and consumption. According to the World Bank, agriculture accounts for 70 percent of all freshwater use.[29] Water use is high due to the types of crops being grown, such as almonds, as well as the inefficiency and lack of incentive for farmers to conserve resources. By some estimates, more than half of the water used for agriculture is wasted due to ineffective irrigation systems, general water mismanagement, and antiquated water rights from decades or even a hundred years ago, when water was plentiful. In the United States, water rights are legally connected with land ownership, and the historic drought in the western United States is leading investors to purchase water rights as a profit-making scheme.[30] In this market-based model for water, water prices, use, and waste are theoretically optimized, but in reality, this has not been the case, demonstrating that these hundred-year-old laws are outdated. In these circumstances, water is considered a paradise for arbitrage, consistent with the energy markets in the 1990s that were exploited by the Enron fiasco.[31] It is yet another example of a market failure to address a critical sustainability challenge.

Because the readily available freshwater resources such as rivers, lakes, and groundwater are overutilized due to economic growth and declining capacities impacted by climate change, 21st-century strategies need to balance the benefits between the public and private good. In some nations, this is as easy as a government decree to take over responsibility or to regulate, which is not necessarily the best option. Property rights make these strategies more nuanced and challenging in other countries, such as the United States. The centuries-old property and water rights laws in the United States are no longer practical given the present-day drought circumstances and must be reformed with a focus on new objectives. Real efforts must be undertaken to improve irrigation systems, water reuse, repurposing efficiencies, and economics in developed and developing nations. According to a National Geoscience study, by changing what is planted on existing fields and optimizing results, output could be increased to feed 825 million people and reduce water usage by 10 percent.[32] There are breathtaking opportunities for improvement that are possible as long incentives are put in place to replace the subsidies that squash innovation.

Dramatic technological advances are underway to improve the utilization and management of water systems. Numerous technologies can be used to enhance the future of farming, such as seed gene modification, AI, the Global Positioning System (GPS), blockchain, and even drones. This field of study, known as "precision agriculture," has been around since the 1980s but has not been associated with environmentalism as much as it has been related to improving profitability for farmers. However, we shouldn't be fooled into thinking these technologies can lead to substantial improvements as long as subsidies, policies, and ownership rights are political issues and opportunities remain for profiteers to swoop in. Today's environmental policies often focus on political expediency, technology, and global agreements without discussing whether existing laws and ownership platforms are in the best interests of both public and private enterprises. Reform is necessary to dig deep into what's happening at the community level to balance public and private interests. Politicians must be held accountable for enabling both. That is the moral of the story: improvements in technology need to have a specific intention. If the definition of the human system isn't changing, technology and other innovations won't achieve the intended purpose.

Solution #3: Community-Based Nontraditional Farms

Farming is one of the most crucial sectors in almost every nation, and there are legitimate reasons for using subsidies to support it. In the United States, small family farms account for 21 percent of production; mid-sized and large family farms, 66 percent; and corporate farms,

12 percent.[33] According to the data, US farms are profitable but highly reliant on public subsidies and an integrated supply chain from farm to table. But if every nation's agricultural system is subsidized, it deters the innovation necessary for environmental and national self-reliance and resiliency.

The weaknesses in conventional farming relative to economics and sustainability are related to its inefficiency in water use and recycling, suboptimal yields per land use (despite improvements), a dependency on chemicals to meet production goals, waste across the end-to-end supply chains, a dangerously high use of pesticides and unsustainable monoculture practices, and the disruption to existing fauna and flora in the wild. After a culture of process improvement within the current model, technologies must be implemented to address further opportunities, and alternatives to traditional farming should be incentivized as well. The broad category of nontraditional agriculture includes hydroponics, aquaponics, aeroponics, vertical farming indoors, and repurposing refurbished sites such as abandoned mines for agriculture. According to Fortune Business Insights, the global vertical farming industry is anticipated to reach almost $21 billion in 2029, up from $3.47 billion in 2021.[34] A 25.9 percent compounded annual growth rate is significant in today's global agriculture market, estimated to be between $3 trillion and $5 trillion, meaning it will grow to approximately 1 percent of the total market. The good news is that vertical farming is growing as a market alternative to traditional agriculture; the bad news is that the pace of growth isn't material to the overall market, and innovations are taking place in the regions of the world where significant food availability exists, not where it is needed.

As an alternative to traditional farming, these methods are an engineer's dream. In contrast to open-field agriculture, which has existed virtually unchanged for ten thousand years, vertical farming modifies and controls nearly every variable—light, temperature, water, seeds, and nutrients—through technology; variables that are important to leave out, such as pests and force majeure, can be eliminated. As a result, vertical farms use up to 95 percent less water and fifty times less land; do not require pesticides; can be set up anywhere, under any weather conditions; and have a smaller transportation footprint.[35] In addition, we are continuing to learn what we know and don't know about growing plants, something we've done for ten thousand years—for instance, the importance of wind to develop strong stalks, the difficulty in replicating the pollination process, and the role of the sun as more than an energy source, given how photosynthesis has evolved over billions of years. We have a lot to learn, and we need to incentivize this process.

Energy is the one major drawback in growing food under these controlled conditions; agriculture has benefited from the sun as an unlimited renewable energy source for hundreds of millions of years. However, photosynthesis is inefficient, with only 1 percent of the light from the sun being used by the plant.[36] Researchers are studying an electrocatalytic process to convert carbon dioxide, electricity, and water into acetate that plants can use to grow in the dark. The most significant challenges in these vertical farms are the high upfront costs and the amount of energy required to run these operations. Of course, any new technology requires higher prices on the front end, and policy should incentivize or subsidize these innovations. The greatest challenge associated with these controlled agricultural processes in community-based supply chains is energy that needs to be achieved through renewable sources, such as solar, and innovations that require less or more efficient processes, such as a replacement of biological photosynthesis. Vertical farming requires us to improve our understanding of the role of sunlight in the plant-growing process.

A community-based supply chain system would be a game changer in addressing poverty and climate change issues, particularly in regions where agriculture is most challenging and those neglected by today's global supply chains. Despite being a small nation with a suboptimal climate, the Netherlands is currently the second largest exporter of agricultural products, behind only the United States, in the world. The Dutch are world leaders in increasing production growth while at the same time focusing on reductions in energy, pesticide use, and other impacts on the planet; they are working on reducing the amount of energy needed to grow produce and have already reduced water use, growing a pound of tomatoes with a half gallon of water, compared to the global average of twenty-eight gallons.[37] Imagine a global food system model that networks more efficient global agricultural supply chains with these localized community farms. In this model, fewer water and energy inputs on less land would lead to higher production output that is more fairly distributed, with lower waste and CO_2 emissions in logistics. There are legitimate paths to this optimized food model if our human systems will focus our strategies in this direction. This is our most significant challenge.

As is the case with our primitive fossil fuel energy systems, so is the outdated nature of the agricultural system. Of course, there is nothing more efficient than growing plants using natural sunlight to achieve photosynthesis, but the other ingredients necessary, such as water and fertilization, are the problem. Much like the energy problem, the technologies are in place to move forward, we just need the human systems to propel them into widespread use.

Solution #4: Plant-Based Dairy and Cultivated Meat

Of the problems that humanity faces, the increase in meat and dairy consumption may be the greatest. According to UN FAO data, meat production and consumption have quadrupled in the fifty years preceding 2020 and will continue to grow. Other data indicate that meat production accounts for approximately 54 percent of the planet's habitable land and 80 percent of agricultural land, but provides only 20 percent of the world's calories.[38] As the worst of the worst, beef cattle consume approximately 25 calories for each edible calorie produced. Meat is a threat to the earth in terms of greenhouse gas emissions, water usage, deforestation, overuse of antibiotics, the ethical treatment of animals, and a risk factor as a potential source of viral infection that may lead to future pandemics. In 2022–2023, the world faced one of the worst outbreaks of avian influenza in history. A vegan diet, completely free from meat and animal by-products, can be a viable alternative and offer health benefits when correctly implemented and can be satisfying from a taste, ethics, and sustainability standpoint. But not everyone wishes to undertake what can be a more challenging path to meeting their daily protein needs, so a strategy for achieving sustainable meat alternatives needs to be pursued rapidly.

Vegans may be correct that humans could transition to a 100 percent plant-based diet to combat climate change. However, this goes against millions of years of human evolution that led to our growth and dominance as a species. The human brain is an energy hog, consuming 20 percent of the body's energy despite being only 2 percent of its weight; protein nutrition is a key need for us to function.[39] Today, our food systems are sufficiently complex and robust to get our bodies the 2,000 to 3,000 calories needed to perform these functions. But historically, this wasn't the case, and meat was the key to our evolution and growth through the ages.

Today, meat consumption has become cultural, embedded in rituals, customs, events, and tastes, but it is also a symbol of status as developing nations become developed and mimic the world's more affluent countries. When I have this discussion in my classes, few students are willing to eliminate meat from their diets, even after learning of its horrifying impact on the planet—unsurprising, since the United States has the world's highest meat consumption per capita, at 219 pounds a year, or 0.6 pound per day.[40] While it may be functionally possible to eliminate meat from the diet of all humans on the planet and provide sufficient nourishment, the human systems are too strong of a pushback.

Plant-based proteins are becoming popular, especially in more prosperous nations, as consumers seek healthier alternatives to meat and dairy products. As a result, the global market of these products was valued at $12.2 billion in 2022 and is projected to reach $17.4 billion by 2027, a 7.3 percent compounded annual growth rate.[41] Given the size of the global meat market at over $900 billion, soon to be reaching a trillion dollars, these plant-based proteins are no threat and not viewed as an alternative to the traditional market. Companies like Beyond Meat and Impossible are struggling, with sales expectations that rose faster than actual sales, and the overall category declined 8 percent in 2022.[42] Companies that produce plant-based meat alternatives will continue to innovate, creating new and improved meat and dairy products. Still, based on current results, it seems unlikely these will displace traditional meat consumption in the Global North or slow down the rapid growth of meat production in developing nations.

Dairy and egg products are staples in our lives that cause similar environmental problems as meat. Megadairies and egg production facilities cause harm to the environment through avian influenza, animal waste, methane release, high antibiotic and growth hormone use, and animal cruelty. Sales of plant-based ice creams made from soy, nut, coconut, and oat milk are growing, as they taste much closer to cow-milk products. In addition, precision fermentation innovations are being developed that use whey protein to imitate dairy products. Dairy-free, plant-based mozzarella cheese can be made using precision fermentation to create casein, the protein in cow's milk.[43] And San Francisco–based Clara Foods is using precision fermentation to create the first animal-free egg product.[44] Of course, all of these companies face scalability challenges. Still, as someone who has worked in the beer industry for a long time and understands fermentation, I believe the process required to achieve viable alternatives to dairy can happen quickly.

The greatest challenge is in a like-for-like replacement of meat, not plant-based proteins where heme is created through precision fermentation to make a burger bleed like the real thing, but the actual real thing! Essentially, the process begins with either undifferentiated and repurposed stem cells or specific stem cells designed for a particular end product. Next, these cells are added to a growth medium in a fermentation tank to proliferate. Finally, nutrients such as fats, proteins, and salts are added to the tank. The challenge is to replicate the way cells grow into tissue in life within these tanks. Once the cells are generated and muscle fiber is created, the product is configured or scaffolded into its intended product.

As with vertical farming, there are significant challenges relative to cultivated meat. Still, when considering these hurdles, we must remember the current problems within meat supply chains. One challenge is the difference between the process of an animal being born, raised, and slaughtered into meat, and meat developed using a cultured process in a fermentation vat. In some cases, these differences are significant, such as the lack of antibiotics, growth hormones, and animal cruelty. At the same time, cultivated meat lacks natural muscle development, and we do not yet understand its impact on the end product. As with plant-based protein, it needs to taste like the real thing, not like some lab creation. Technologies must improve, for example, to replicate how the animal development process impacts flavor. The next issue is concern around genetic modification, with the potential for dangerous changes to the food supply chain that we do not yet understand. Yet this concern may be no more problematic than the dangers associated with current meat production, which have and will continue to lead to animal-to-human viral transmission, and other supply chain issues from farm to table. And finally, perhaps the most pressing matter is the cost of the product. Prices of plant-based meat products are falling, but it is questionable whether they will eventually reach parity with traditional meat. In a 2022 study published in the *Journal of Agriculture and Food Research*, the authors found that even if cultured meat becomes scalable, the end cost would be $63 per kilogram (approximately $29 a pound) and a hamburger would cost $18.[45]

Even if these challenges are addressed, and I think they will be, the most significant challenge will be in our human systems: consumer adoption. One food tech start-up, Future Meat Technologies, estimates that a third of US consumers would adopt cultured meat, hardly a ringing endorsement.[46] I get a similar response when I ask my students how many are ready to embrace cultured meat. Yet when I ask if any of them understand the current Frankensteinian nature of our industrialized factory farms, very few of them raise a hand. Cognitive dissonance runs rampant among those who profess to be concerned about climate change, understand the nature of the existing supply chains, and are reluctant or apathetic about new solutions. In so-called climate-change-conscious regions of the world, such as Europe, concerns relate to synthetic biology. There are barriers to overcome relating to the future of farming and its impact on farmers, but shouldn't government policymakers and consumers have even more significant concerns about the existing supply chain systems? Our refusal to acknowledge the frightening nature of our current state reveals the embedded nature of our cognitive dissonance. If we are serious about supporting the environment, changes will need to be made in our attitude, our culture, and even our diet.

Solution #5: Community-Based Water Systems

There is undoubtedly a water crisis worldwide. According to one 2021 study, more than half of all the world's rivers run dry at least once a year, including some of the most important, such as the Nile in Africa, the Yellow in China, and the Colorado in the United States.[47] China is facing its worst water crisis on record, particularly in the south, which could significantly impact its economic development.[48] The same is true for nearly every fast-growing developing economy such as India, and it is at least a partial flashpoint in regions such as Iran, which is in overdraft for most of its groundwater resources, which could lead to political instability. Pakistan is facing record droughts and floods concurrently and is a large nation in peril. In the United States, the western reservoirs were drying out but are now rising; Lake Mead was dangerously low at 28 percent capacity and Lake Powell at 27 percent, but both have since recovered through fortunate rain and snowfall and water conservation efforts. Yet the American Southwest is facing the most significant drought it has seen in twelve hundred years. In Africa, where most farmers have no irrigation systems and therefore rely solely on rainfall, droughts are a catastrophe.

In October 2022, the US Geological Survey (USGS) revamped the water cycle diagram we learned as kids—evaporation, condensation, cloud formation, and precipitation—to account for the anthropogenic element in the process.[49] Across the planet, there isn't one water cycle, as some of us have learned, but a competing and conflicting array of overuse and lack of effective management. Instead, what should take place is what I've discussed: a set of community-based systems acting in a network. Engineering fixes to the present state, such as dams and canals in rivers, are band-aids that address the symptoms rather than the root causes of the problem. For instance, these engineering solutions address the lack of water without addressing the reason for it: the industrial misuse through our human systems of supply chains. A lack of strategy related to human systems in cooperation and collaboration through a networked community-based approach needs to be addressed.

Another issue embodied in our water policies, rights, and laws is our tribal nature as humans. These policies, rights, and laws are perplexing, differing from nation to nation, state to state, and locality to locality, translating into a Tower of Babel disaster that needs to be addressed in a 21st-century sustainability model. The model should include freshwater, groundwater, wastewater, brackish water, and saline water in a comprehensive strategy and solution. Treated and recycled wastewater is as clean, if not more sanitary, than conventional water samples and should be expanded across the planet. The same possibilities exist for other types of water. An example of improved water use highly related to agriculture is China's growing implementation of "seawater rice," which doubled its yield in three years.[50]

Community-based systems must use existing water resources more efficiently, focusing on rainwater collection, waste reduction in residences and industry, and leak management. Beyond these preventative measures, which will make a significant difference, new sources such as brackish, salinated, and nonconventional uses of nonsalinated water are introduced into the water stream through technology. The proposed solutions must be evaluated and implemented on a community-by-community basis, based on their circumstances. Networking communities to flow water from a supply-and-demand standpoint, as I have also proposed for energy, will also be necessary because it isn't just a supply issue, but a distribution issue as well. If these efforts are undertaken, there will be more than enough water to meet the world's needs.

There are so many game-changing opportunities to create a 21st-century model for food and water if today's policymakers and environmentalists focus on finding solutions based on data and analysis of the situation rather than obsolete laws and political benefit. We can solve the imbalances in the food production and water systems through emerging technologies and thoughtful considerations of how the supply chain distribution systems can be networked and locally resilient. We have overcome the concerns of 18th-century economist Thomas Malthus and can grow more than enough food for eight billion world inhabitants. Our planet has more than enough water when we redefine the term. Challenges exist for how we need to feed the world and provide clean water, but differently than how we consider these problems to exist today. We live in an exciting time of new possibilities for solving old problems!

Chapter 8

We Are Living in a Material World

The late 1970s and early 1980s were a rough time in American history. Inflation in 1979 was the highest on record, at 13.3 percent, and unemployment rose to 8.5 percent in 1981, with nearly a quarter of all autoworkers not working. When Ronald Reagan became president in 1980, he put the blame for the economy's problems squarely on the government, promising he would get it off the backs of the American public so they could live in economic liberty. Reagan delivered on this promise, cutting the government's role—and control—and ushering in an era of unprecedented consumerism. During the Reagan era, the US economy launched into the longest peacetime expansion on record. As a result of deregulation, big businesses got bigger through leveraged buyouts, and in the 1987 movie *Wall Street*, corporate raider Gordon Gekko coined the now famous phrase, "Greed is good." Madonna's song "Material Girl" was released in 1984 as a satirical take on consumerism in society, but it was taken seriously by the many young women who saw her as a role model. Consumption grew in the United States after World War II to save the economy from the collapse in its transition from war, and since then, there has been a downslide away from balance and toward a model focused on consumption. In the 1980s, the model was perfected. Madonna was correct that we are living in a material world—and the planet is paying the price.

In truth, humans have always lived in a material world, but with different connotations. Even before humans, primates used materials and tools, and our most recent ancestors used stones for protection, hunting, chopping down trees, and the like. Human survival and evolution, unlike that of any other animal, depended on technology, starting with simple items made from nature and fire, which were essential for many reasons. Pottery made from clay some fourteen thousand years ago was helpful for water and food storage and consumption. Around ten thousand years ago, critical discoveries were made in the use of metals such as gold, silver, and copper for cooking, weapons, and ornaments like jewelry, but it wasn't until five thousand years ago that the metal alloy of bronze was developed, leading to more substantial and durable products. Subsequent metals such as iron and lead became valuable as human society advanced; lead was allegedly one of the causes of the fall of the Roman Empire, poisoning the Romans and driving them crazy. Cultures and great civilizations advanced using materials, but it was never enough; demand and functional needs always exceeded supply. For thousands of years, humans living in a material world weren't materialistic but in need.

Earth is a closed system except for the rare asteroid, so the total mass of materials doesn't change, only their accessibility. For thousands of years, some of these materials were insignificant to humans until new knowledge and technologies allowed us to convert them into something we could use. For example, bronze was alloyed from the metals copper and tin some five thousand years ago, but then there were supply issues around 1200 BCE, leading to the transition to iron. Yet it wasn't until the 13th century, when steel, an alloy of iron and carbon, became useful, and not until Henry Bessemer's process in 1855 that its supply could scale. Aluminum is another example; rare in its natural form, it was more valuable than silver and gold and used to cap the Washington Monument in 1884. Once the methods were perfected to extract alumina from bauxite and convert it into aluminum using high energy, aluminum became an inexpensive, bountiful commodity. Scientific knowledge, energy utilization, and supply chains have enabled today's material world; the question moving forward is whether use of materials in industry in balance with nature is possible.

Modern Material Flow

To understand the scope of the material world, I start with the 2022 *Circularity Gap Report* produced by Circle Economy.[1] This report estimates the global material flow, measured in gigatons of material extracted, processed, produced, and consumed, and its end-of-use state after that. Given the private, complex, and global nature of today's supply chains, it is difficult, if not impossible, to track materials from raw to in-process to finished goods to post-use. Better systems can be implemented to track and manage materials. Part of the challenge is a near-infinite definition of the term "material." To solve a problem, we must first understand it, and given the number of permutations of material elements, components, compositions, and so on, there must be a better understanding moving forward to address our environmental woes.

According to the Circle Economy report, the global supply chains require approximately 100.6 gigatons of material, including 92 gigatons of newly extracted materials and 8.6 gigatons of recycled material, a less than 10 percent reuse rate. Of the raw materials for the production source, 50.8 gigatons are minerals, 24.6 gigatons are biomass, 15.1 gigatons are fossil fuels, and 10.1 gigatons are ores. The end uses of these converted raw materials are as follows: 38.8 gigatons for housing, 21.3 gigatons for nutrition, 10 gigatons for services, 9.3 gigatons for healthcare, 8.7 gigatons for mobility, 6.9 gigatons for consumables, and 5.6 gigatons for communications. The first thought from these data is that the sheer volume of extracted raw materials requires towers over the recycled/reused materials. In today's supply chain system, extracting raw materials is more market viable than repurposing existing ones, leading to our waste management problem. Such a model cannot last a few decades, much less forever.

Food and shelter are foundational in the hierarchy of needs, so it isn't surprising that it comprises 60 percent of materials. Many of these materials are organic and carbon based, primarily renewable, and need to be grown and raised. In the mainstream focus on sustainability, there is much emphasis on the transportation sector, given its energy consumption. Still, it is often a secondary factor in the supply chain, such as its use in food production and distribution. In this report, the growth of materials used in the 21st-century economy is becoming evident in newer technologies (communications and services) and in healthcare to serve the aging population. However, what remains is often the most disposable of materials and consumables, only 6 percent of material use but a much higher percentage of waste. While materials science is a field of study, it primarily focuses on the invention, production, and use of materials; it must increasingly focus on reusing materials for economic growth and sustaining the planet.

The global material flow analysis conducted by Circle Economy is a significant undertaking but only a high-level estimate of today's complex, decentralized supply chains that span 195 nations. There is no way of knowing the exact flow of raw materials through inventory, production, distribution, retail, use, reuse, recycled materials, and waste. Each raw material category is broad, such as minerals, estimated as 55 percent of total material use spread across anywhere between four thousand and ten thousand minerals on earth.[2] Most of these minerals are rare, and in industry, most of the mineral use is in a much smaller number of materials, including limestone, clay, sand, gravel, silica, and others. From there, these substances become part of various building materials, paint, ceramic, glass, plastic, paper, electronics, detergents, medication, medical devices, and many others.[3] The same levels of complexity are the case with ores, biomass, and even fossil fuels (not only oil, gas, and coal but also many synthetic materials) and other supplementary uses. In addition, there is no record of how extracted materials are used in private enterprise, as metals are alloyed and other materials become composites in virtually unlimited combinations. There is no question that capitalism and private enterprise have led to immense innovations and technologies that have improved society, but from a planetary standpoint, it isn't easy to manage the impact. These resources are not unlimited, and their poor use and management will lead to challenges. Additionally, improper use and overuse harm the health of the earth. For humans to understand and better manage the planet's health, we must better understand how to harness, harvest, grow, use, consume, and reuse the world's finite materials.

Human understanding and management of materials need to shift from simple acquisition, conversion, and use to acknowledging that we hold dominion over them on this planet. We should treat this responsibility much differently than we do today. However, because we are tribal creatures, we often consider acquisition of materials a competition, given their finite nature, leading to a lack of cooperation across nations and public and private enterprises for their industrial use. There is also a "use it or lose it" mentality related to water, oil, and other commodities that is dangerous to the environment.

In contrast to agriculture, where the wealthiest countries often have the most significant access, raw commodities such as oil, gas, metals, and minerals are often found in poorer regions, making material extraction crucial to the economy of those regions. Many of the largest petroleum and mineral exporters are developing nations in Africa, Asia, and South America, which face the so-called paradox of plenty, where wealth has led to inequality and corruption. In Mexico, drug cartels are working their way into the lucrative lithium-mining business. Latin American nations such as Bolivia, Argentina, and Chile are discussing setting up a possible OPEC-style cartel for lithium, as these nations own 63 percent of the world's reserves.[4] Developing African nations, such as the Democratic Republic of the Congo (DRC) and Guinea, are superpowers for these 21st-century minerals, and the military takeover in Niger has raised geopolitical tension, given its huge uranium deposits. In too many cases, the general population of these nations does not benefit from the economic gains.[5] The balance of power is tilting in the direction of these developing nations, which must become responsible for transforming this into equitable economic growth. For example, the DRC supplies 70 percent of the world's cobalt, but Chinese companies own or finance 80 percent of these mines, an example of a lack of self-reliance.[6] In a study of eighteen critical minerals found in South America, sub-Saharan Africa, and Southeast Asia, nearly all dealt with middle to high levels fragility and corruption, with Australia as the only exception.[7] The average consumer is only faintly aware of the unsavory and unsustainable nature of these raw commodity markets. In today's complex global supply chains, the end consumer is rarely aware of the circumstances behind their smartphones and other purchases.

Global commodity markets will never lead to a global strategy for managing the planet's natural resources and providing equity and prosperity through their use. Throughout history, commodities have been the key to civilization and often the cause of wars, revolutions, and other conflicts. Collectively, we live on a finite planet, but the utilization and management of the millions of natural and manufactured materials have few controls. As a result, we face a global crisis that requires a different approach to handle an important topic that has become a 21st-century challenge. Otherwise, humans will ravage our way through finite commodities while destroying the planet.

Solution #1: Bill of Materials Documentation for All Products

There are 195 nations in the world, each with different policies and laws, and many large, powerful multinational corporations that answer to no nation. There have been valiant efforts, mainly in Europe and North America, to hold manufacturers responsible for the sustainability of their products, with extended producer responsibility laws. There have also been laws that hold the consumer accountable for the sustainability of the end-use product, including bottle bills in the United States and Europe that require consumers to make a small deposit on bottles and cans, that is refunded once they are returned to a recycling center. I have been studying solutions to the reuse problem for more than a decade, and my book *The Recycling Myth*, published in 2015, questioned the effectiveness of these policies. When I was writing that book, I tried unsuscessfully to convince environmental groups such as Greenpeace to study the data that show these policies to be ineffective. In 2022, Greenpeace conducted its own study and came to the same conclusion using the same terminology: recycling is a myth that doesn't work.[8] The science and math around recycling some materials, including plastic, don't work. The most significant challenge is that companies are not required to provide a detailed bill of materials (BOM) for the product sold in these supply chains. For food products we consume, companies are required to provide a list of ingredients, but the same is not true for any other materials in the global supply chain.

The root-cause problem is a supply chain challenge: the manufacturer, known in the business as the original equipment manufacturer (OEM), often doesn't know the exact chemical, physical, and electronic properties of the alloys, polymers, and components in the product because there is little transparency from one end to the other of the supply chain system. Take, for example, a plastic bottle made from a base polymer of PET. The manufacturer understands the base polymer but not the additives used to produce the bottle to its required specifications; therefore, neither the producer of the bottle nor the soft drink manufacturer can provide information to the consumer regarding the materials used. Likewise, an airliner is said to have millions of separate components on its BOM, and it is not the responsibility of Boeing to account for the specific materials and chemicals used in the production of the plane. The smartphone you're using is assembled in China, but approximately forty different minerals from other nations around world comprise components made in other countries. Apple, Samsung, and other manufacturers keep few records of these detailed BOMs, critical information for material reuse, repair, and even ethical business to avoid conflict minerals. We have to ask ourselves how it can be possible to better manage the complexities of the near-unlimited scope and scale of materials through extraction, production, assembly, use, and post-use without supply chain transparency.

Why don't detailed BOMs flow from the beginning to the end of the supply chain? The answer is: nobody is responsible, the process would be too complex, and there is no single platform to track these materials. Essentially, it is a people, process, and technology problem. From a process and technology standpoint, emerging technologies can solve these challenges, with the most notable being blockchain. Blockchain is a centralized-decentralized ledger system that can provide rights to users based on what information they need within the process. In this case, the manufacturer of the material could be held responsible for documenting the detailed BOMs for all the chemicals, materials, and energy used to make its product. Granted, setting up a system like this will take a lot of work and coordination worldwide, but these sorts of coordination efforts have been done in the past, such as the modern twenty- or forty-foot shipping container, which moves from one location to another seamlessly through standardization of design. In addition, humans know how to tackle complex challenges when incentivized within their communities. As is the case with other challenges I've discussed in this book, the greatest challenge won't be technology, but the human systems coordination.

Today's global supply chains are very fluid and complex and lack any sovereignty that requires one supplier from one country to follow the rules of consumers in another. According to World Bank statistics, the world wastes 2 billion tons of materials annually, which is probably understated. This number is projected to increase to 3.4 billion tons by 2050, with at least a third of it not properly managed.[9] These data show high growth rates in waste by 2050 in the developing regions of sub-Saharan Africa, South and East Asia, and other areas without formalized waste management systems. Waste within the privatized supply chains is considered an externality, and except for a few wealthier nations with smaller populations, there are few incentives to address these challenges. There are increases in dry waste in more affluent countries, such as single-use plastic, nondurable goods (products intended to last less than three years), and electronic waste, much of which is cheaper to throw into a landfill than to recycle. There is more organic waste in poorer nations, and less sophisticated waste is often dumped without proper precautions.

Modern supply chains are global in that one company can receive an order in the United States, manage inbound materials for production, produce the product, and then ship it from their factory in Asia to anywhere in the world. From a financial and economic standpoint, this is as seamless as possible, which has led to significant growth over the past century. But this process doesn't reconcile with the limitations of resource availability and the environmental impact. Without a strategy to document the materials as they flow seamlessly across the planet, there will never be sufficient information to reduce the waste that is so prevalent today. If materials aren't tracked and traced throughout the supply chain, how can we manage their efficient use, reuse, and waste? If policymakers are serious about reducing waste and better managing the limited resources on the planet, a new approach to supply chain must be adopted; as someone who spent decades in the field, I know this is possible if there's a will to do better.

Due to property rights and self-interest, private corporations and governments may push back on implementing reforms and innovations such as a blockchain system. For example, multinational corporations and their suppliers may argue that providing a detailed BOM for their products requires too much work, adds cost to the end product, and will compromise their intellectual property. These are the arguments of the 20th-century supply chain that can be addressed by reinventing a supply chain for the 21st century. It's as easy as documenting your materials, coordinating, and collaborating through advanced technology. Proprietary formulations and designs can be protected through the blockchain; it will be feasible for a manufacturer to own the centralized design of the product and other rights based on the ownership of the materials. Given today's complex global supply chains, establishing an end-to-end documentation strategy for development will alter the legal definitions of "property" and "ownership." Therefore, we need to develop a system for the public good that also protects private enterprise.

National security and self-interest present another roadblock to overcome, particularly given the nature of geopolitics today. As such, some opportunities, such as microchips, may be off-limits or more closed systems due to national security concerns, but most material flows shouldn't face such scrutiny. Supply chain transparency must play a role in developing new products and technologies and procuring critical metals and minerals for economic growth and environmental sustainability. Requirements, or even guidelines, for 195 nations to support a "design for supply chain mode" to require manufacturers to document their BOMs don't have to be an affront to intellectual property and ownership rights if a blockchain system can be used enable security and privileges based on roles. This system could be used to improve the environment and the rights of individuals and, at the same time, secure the intellectual property of the owners in the supply chain. This strategy won't lead to an increase in product cost; it would provide the opportunity to reduce cost through coordination within the supply chain system. Studies have found that if the cost of environmental sustainability is reasonable, consumers will support it. This initiative would serve the best interests of the consumer through improved supply chains and transparency for environmental action.

Finally, this end-to-end BOM documentation is an applied approach to managing material waste and emissions that is in contrast to methods such as carbon accounting that only track and don't enable improvements. Also, BOM documentation provides the BOM for *all* materials, not just fossil fuels. It can be used at the end of a product's life for mitigation, redesign, and waste management in the supply chain. Carbon accounting tracks only carbon dioxide, but there are many other materials that are just as essential to track. In contrast, a design for the supply chain model, which is designing a product from "cradle to cradle," can be the first step in allowing us to manage finite materials on a finite planet. The starting point, and its use in blockchain, will be wonky and take time, but we cannot solve the problems facing our environment without undertaking this challenge.

Solution #2: Material Genome (Design for Sustainability)

Another reason it's crucial to track, trace, and manage the detailed BOM of products produced and used worldwide is to measure and manage materials to make them better for the environment. I teach my students in supply chain and sustainability classes that products aren't designed to be sustainable, and in today's outdated supply chain system, they are made in a market system that focuses on short-term profit over economic and environmental sustainability. So, first, supply chains need to document the data in materials using a secure, high-technology blockchain solution and then be responsible for improving upon it in designing the product for sustainability. Emerging technologies can enable this, but before focusing on technology, we need to address the root cause problem: our human systems. If we don't, then the technology will not be of good use.

The sequencing of DNA began in the 1970s, and by the late 1970s, the first DNA genome—bacteria—was sequenced. Human genome sequencing was officially completed in 2003, with 92 percent of the job done, and the final 8 percent, crucial to understanding our biology, was finished in April 2022.[10] Twenty years after its official completion, the Human Genome Project has led to breakthroughs in the diagnosis, early detection, and treatment of diseases, including a central role in developing a COVID-19 vaccine, deployed less than a year after the sequencing of the virus. And yet, if you look at the research and development process, it's taken decades in scientific development, and then added time for applications to improve healthcare and other fields. In 2011, the White House, under the leadership of President Barack Obama, launched the Materials Genome Initiative (MGI) to help businesses discover, develop, and deploy materials twice as fast.[11] Similar initiatives are happening elsewhere, such as the National Materials Scientific Data Sharing Network in China and similar efforts in the EU. The concept of MGI is analogous to that of the Human Genome Project: to establish materials data in a standard format in an open-source manner to enable scientists, engineers, and entrepreneurs to redesign products in the supply chain.

Progress happens in material genomics when new materials are sequenced and developed. Studies are conducted to determine how new materials, either those already in existence or completely new creations, perform necessary functions, such as a proper storage tank for hydrogen use in vehicles. These material databases are combined with AI to develop new nanoparticles; in one experiment, researchers ran a model to create new materials from a mixture of seven elements, and the computer generated eighteen new possibilities.[12] These newly developed materials, called metamaterials, have properties not found in naturally occurring materials. In contrast to designing novel structures from existing base materials, metamaterials are entirely newly designed structures that respond to natural phenomena such as sound waves or radiation in a preferred manner. Today's conventional materials are limited to finding existing matter and manipulating it to meet specific requirements through chemical, reductive, or additive manufacturing; metamaterials are a paradigm shift outside of this current realm.

Material genomics, metamaterials, and nanomaterials are inextricably linked as concepts relating to the future of materials. Nanomaterials are defined by the size of the material, between 1 and 100 nanometers. To give you a sense of scale, the period at the end of this sentence equals 500,000 nanometers, so the largest nanomaterial is five thousand times smaller than the period. Nanomaterials can be naturally organic, inorganic in nature, or metamaterials formed from processes. The benefits of nanomaterials related to improving the economy and the environment are enormous, including water purification, sustainable agriculture through growth management, replacement of pesticides and other harmful elements, waste management, and energy transformation. From this perspective, materials are defined more by the requirements of a material than by their existing organic, natural, inorganic, or synthetic state. We are about to witness a game-changing era for how the human-made can interface with the natural world.

One material that is in significant need of redesign is concrete, which is responsible for 8 percent of the world's carbon emissions. Concrete comprises sand, gravel, stone, and cement, made by roasting limestone. The creation of this material is very intensive from an end-to-end supply chain perspective. Not only does it take a great deal of energy to create, but in its afterlife, it has a low recycling rate, as low as 30 percent, due to contaminants and the lack of structural integrity of the end material.[13] Concrete is a current-state problem, as there is legislation in the EU, the United States, and other regions to make concrete more sustainable in the absence of sufficient alternatives. There have been efforts and some progress made, such as using graphene, a thin carbon layer, an as additive that allows for a product to be stronger through less material, reducing concrete use and waste. There are also efforts to reduce the amount of energy used to produce concrete by using bacteria to create energy using light rather than heat. Using "ancient technology," such as bacteria, which have existed for billions of years, or even techniques from Ancient Roman concrete production can help us look at design much differently. The efforts underway are in the beginning stages, but excellent progress is being made in tackling the environmental challenges posed by concrete. Concrete is an example of a product that hasn't been redesigned for decades; through material genomics, new solutions are now possible!

When renewable energy replaces fossil fuels, there will need to be suitable feedstock alternatives for manufacturing many materials, including chemicals and plastics. According to the US Department of Energy, more than six thousand everyday products require fossil fuels as a feedstock, and this doesn't include by-products from the petrochemical manufacturing process, such as CO_2, which is used to carbonate beer and soda and in food production and medical facilities. Approximately 7 percent of the world's fossil fuel consumption is in noncombustible means, with 4 percent used for plastics and the remaining for chemicals and additives in a near infinite array of products. Fossil fuel is the base polymer for virtually all of the plastic in the world, a material that is recycled at a low rate in total, less than 10 percent, and ending up in our oceans and being consumed by marine life but would be impossible to do without in our daily lives. Bioplastics, which have an organic feedstock such as corn, are a poor substitute because of their chemical composition, manufacturing process that diverts resources that could be used for food, and overall worse environmental impact. Numerous chemicals are added to plastic to make it softer or harder, gas permeable, flame retardant, inflexible or flexible, and so on. New materials, conversion processes, and energy sources (e.g., light versus heat) to create plastics are required to meet our environmental goals.

As noted earlier in this chapter, there are no classifications of specifications, recipes, and so forth that categorize chemicals, plastics, and other synthetic materials, and much work is needed to be done if we are to replace these plastics and chemicals in the world. Because there is no official documentation for every plastic material formulation invented and deployed worldwide, we can only estimate that there are more than hundreds of thousands of chemicals used in industry, with most of them not documented and very few regulated in industrial and consumer use. Given this lack of documentation of the chemicals in use, nobody knows of their impact on the environment and our bodies. Yet today there is an increasing number of likely related human diseases and environmental impacts that we are experiencing without a proper understanding. It is an almost unfathomable problem to diagnose today, when so few synthetic materials are documented and little is disclosed related to their formulas.

Material genomics using advanced technologies such as blockchain, AI, metamaterials, and nanotechnology/nanomaterials is our greatest hope in balancing environmental sustainability and the economy. Much effort has been put into developing future energy options, and there has been much progress; the same is true for food production and water availability. But there has been an insufficient focus on the role of materials and how much energy is used to create them. Energy, an essential feedstock beyond these most apparent uses, must be reimagined. The problem with materials is that there are millions of organic, natural, inorganic, and synthetic materials in the world, with few records to track, trace, and manage them. Understanding and digitizing material definitions must be the first step, and then we need to use advanced technologies in material genomics to improve and replace them for the 21st century. More effort needs to be focused on this space.

Solution #3: Design for Supply Chain/ Community-Based Systems

These first two solutions presented in this chapter are transformational improvements to the definition and design of materials, but they wouldn't necessarily improve how today's industrial supply chains operate. Today, materials are designed, procured, and produced based on product specifications, but they are not designed for end-to-end use after their useful life. Therefore, a material that is 100 percent renewable, such as aluminum, has only a 76 percent worldwide recycling rate, and the rate is much lower in the United States despite aluminum being efficient and infinitely reusable.[14] Of course, aluminum has an exceptionally higher recycling rate worldwide than plastic, at 9 percent. Still, the former is an efficient reuse material, while the latter is not.[15] Both materials provide insight into what's wrong with today's supply chain: aluminum's supply chain is broken because a completely sustainable material is being recycled and reused at too low a rate, and plastic is not designed for reuse within the end-to-end supply chain. Therefore, simply implementing Solutions #1 and #2 will not necessarily improve the extraction, conversion, distribution, production, and reuse of materials in the industrial supply chains and consumer markets and improve environmental sustainability. We must also redesign our supply chains to work more effectively.

I use a concept that I call "design for supply chain" with my students to explain how products should be designed from a material standpoint through the end of the consumer market process. Solutions #1 and #2 address how existing and new materials must be designed and redesigned to improve sustainability. This third concept ensures that both the material and the end product can be effectively managed as a material on earth beyond its consumer use. This holistic approach, which manufacturers must be responsible for undertaking, combines what's best for both the environment and the consumer. Unfortunately, very few manufacturers are considering the environmental impacts of their products and the materials used to create them, despite the marketing, greenwashing, and ESG programs that claim to promote the public good. They cannot do so because, under the current market-based system, they need to focus on what the consumers want: the highest-quality, cheapest product possible. The question is, how can we change this model to switch the focus from consumer price to value to the economy and the environment?

In my model, Solution #3 is a community-based supply chain system built on "design for supply chain" and "design for sustainability." Rather than replacing the existing global, long-tailed supply chain system focused on financial markets and waste, I am proposing an alternative that acts as a closed-loop system for materials. Although there is much discussion and many claims made related to a so-called closed-loop system, such a system, by definition, isn't possible within today's supply chain design. While some materials, such as aluminum, are valuable due to their efficiency and ubiquitous use, many materials are not, with an inelastic intrinsic value starting as a feedstock and ending in post-use. Plastics, for example, are commodities fit for a specific type of use based on a fossil fuel commodity where cost is most important and the end-of-life value is limited; therefore, the only economically viable manner to reuse it is to keep the supply chain costs as low as possible, including the logistical costs of shipping to a conversion center. In nearly every case, these materials are not viable for reuse from a supply chain perspective, so they are dumped in landfills even after being "recycled" by consumers. For reuse of many of these materials to be viable, future products must be material genomed and documented with product BOMs. Once this happens, the closed-loop system comes into play, with the post-use material having value because it resides in the same locality as the remanufacturing facility.

If today's long-tailed supply chains are to remain in place, the modern-day challenges of material management must be addressed. The limits of natural resources and environmental impact are new economic variables that must be factored into economic models and financial markets. When we do this, the financials look much different from today's model, especially given the capacity of technology using 3D printing to produce a single item at a cost increasingly competitive with large quantities. Community-based supply chains will become viable alternatives to the existing supply chain system because of new environmental cost factors, a societal need to better balance production and consumption, and new technologies. As technologies continue to improve the use of 3D printing to optimize the resolution of manufactured materials and the speed of the process, the interfacing of manufacturing and materials will naturally become scaled at lower levels, leading to the community-based system as the innovation.

Solution #4: De-Materialization

The concept of "degrowth" has been around as long as the environmental movement, starting in the 1970s. The term was first used in the 1972 report *The Limits to Growth*, commissioned by the Club of Rome, an impressive think tank of business leaders, scientists, and other intellectuals concerned about the state of the planet and our industrial systems. Other academics and think tanks conceptualized similar models based on isolating certain variables and not considering others, such as socioeconomics and poverty, topics I've covered in this book. Degrowth also doesn't factor in evolution, which drives an organism's aspiration toward growth as a matter of survival. Humans have survived and become the dominant species on the planet due to development, and the concept that this will need to change is without merit to us or any organism. The idea of degrowth sounds best from a conceptual standpoint, separate from the realities of evolution.

The problem isn't growth, per se, but material growth; proposing an end of something that is good due to a negative effect isn't logical. Rather, why not address the symptom without discarding what is good? A de-materialization of development can enable growth without a dangerous environmental impact. Therefore, through deductive reasoning, it makes sense that de-materialization is one of the essential concepts for ecological sustainability, because humans cannot and should not try to change our nature toward growth. Therefore, economic growth must be disconnected from materialism. As a result, de-materialization isn't a technology or an innovation but rather a human system that requires us to change how we think, produce, and consume. For nearly eight decades, Americans have been acculturated to consume things made by others. Now, we have to envision ourselves as producers and consumers in a market system that enables growth through using fewer, not more, materials.

There are some basic methods for de-materialization in the supply chain. The first is through the efficient use of materials. For example, the weight of the beverage aluminum can has been reduced by 38 percent since the 1970s, meaning manufacturers are able to do more with less.[16] Commercial airplanes have been de-materializing for decades, which has led to a reduction in fuel use. However, cars are heavier today than they were in the past because vehicles are larger and contain more on-board capabilities and technology; even smaller electric vehicles are heavier due to the weight of the batteries.[17] Therefore, finding opportunities to make products using less material is the best way to de-materialize our markets, but this will require a paradigm shift for product designers, supply chain professionals, and consumers who do not make this a current priority despite all of the discussions happening around environmental sustainability.

The second strategy for de-materialization is improving the recycling and reuse rates of existing materials. When factoring in the use of materials across the supply chain, as is presented in the *Circularity Gap Report*, only 24 percent of materials are reused, with significant variation between those of high and low reuse.[18] As mentioned above, a considerable element of recycling and reuse is the product's design and the overall supply chain, addressed in Solutions #1–3. However, economic models and attitudes among manufacturers, suppliers, and consumers must also change. For one, manufacturers have increased revenue and profitability through waste, such as in the planned obsolescence of products and designs that make them difficult or impossible to repair, such as the Apple iPhone.

Manufacturers have also used expiration date schemes and packaged quantities to increase profits through waste. On the other hand, consumers expect products such as produce, as well as packaging, to be perfect, even when it doesn't impact the integrity of the product inside, leading to higher waste and higher profits. Culturally, we live in a disposable consumerist society in cognitive dissonance with the philosophy of concern for the environment. When I ask people why their actions differ from their views on sustainability, they sometimes scoff or even get offended. Yet we need to understand these apparent contradictions.

A sharing economy will be necessary for the de-materialization solution. For example, most of us utilize our cars only 5 to 10 percent of the day, yet we own rather than share our vehicles. The same can be said for many items in our lives, yet we have an ownership rather than a sharing mentality related to these things. When I raise these questions to my students, many of them scoff at the notion that something shouldn't be at their disposal exactly when they need it, even when I suggest that a future-state model of items as services rather than products could be almost as convenient as owning them is today. Many countries are moving to share models for cars, bikes, scooters, and other items. Using smartphone apps, it is easier and more cost-effective than ever to schedule someone to mow your lawn, drive you to the airport, or fix something in your house, so you don't need to own as many tools. The technology, accessibility, and cost will continue to improve, but based on the conversations I've had with students, friends, and general audiences, the most significant shift will need to happen to our mindsets.

The digitization of consumer products is a significant opportunity for de-materialization. A five-ounce smartphone fits in your hand and has functionality that used to require fifty discrete products that weighed more than seventy-five pounds combined![19] There are some inherent problems associated with de-materialization—smartphones, to continue the example, contain trace amounts of materials in their components that make them difficult to recycle and reuse—but these issues should be addressed by Solutions #1–3. This concept is called "sustainable digitization." Not only is this the making of once-physical items digital, such as a paper ticket for an event or a subway ride replaced by a digital ticket in your smartphone wallet, but also in the use of digital technologies to reduce material use, such as internet of things (IoT) devices that reduce the need for products to perform tasks. In our modern society, we use electronic devices beyond our smartphones for various purposes, meaning there should be an opportunity for fewer things. And yet, despite the potential to design more product capabilities within one device, such as our smartphones, we have more things in our lives than ever before. Once again, technology is available, but our human systems and behaviors must adapt.

De-materialization should mean changes to our supply chains and economies. Supply chains need to be experts in data and information to succeed in a 21st-century economy, being more than focused on trucks and factories; it needs to be an electron-based more than a fossil fuel energy–based economy. De-materialization will be a good thing for reducing energy use when renewable energy powers the electron-based information system. In addition, 3D printers and blockchain technology will enable community-based systems to be networked worldwide, leading to greater sophistication and lower inventory levels, another source of material waste in the supply chain.

Finally, de-materialization should also mean a new human systems model of economics where growth is no longer achieved through materials and the waste of them, such as single-use, throwaway items and planned obsolescence. There is little argument that marketing campaigns have enabled—and continue to enable—development through waste, starting in Western economies and spreading worldwide. Companies and economies must define new strategies without achieving economic growth through consumption and waste. Citizens who say they wish to do something about the environment must consider the cognitive dissonance in their lives. Human systems must design a rational economic and supply chain strategy that doesn't throw out the baby (economic growth) with the bathwater (consumerism) and continue the current state. This will be our greatest challenge.

Solution #5: Redesign Materials on Earth

Finally, Solution #5 acknowledges that we have reached the Anthropocene, the current geological age when activity dominates the climate and the environment. There is no turning back to a different era, and as such, we need to take responsibility for making the planet more habitable under our current circumstances. This ethical conversation is a scary sci-fi scenario in which humans take over control of the planet and play God. Because we are limited in our knowledge and understanding, there are questions about whether a geoengineering strategy will help or make matters worse, and whether we can truly understand until it's too late. Even considering the solutions presented in this and the two preceding chapters, there are questions of whether these 21st-century strategies are enough or, in some sense, humans and the planet have reached an inevitable point of no return.

Geoengineering focuses primarily on removing CO_2 from the atmosphere, using solar radiation to cool the planet, and seeding the clouds to enable rain. Still, numerous opportunities exist to repair the planet. One example is the use of genetics and germs to create bleach-resistant coral to prevent the death of coral reefs.[20] Other geoengineering applications are global tree-planting efforts (afforestation), carbon sequestration, ocean fertilization, and ocean alkalization (to prevent ocean acidification). Some methods, such as sulfate spraying in the skies to mimic volcanic activity as a means to cool the planet, are controversial, unproven, and might lead to problems and geopolitical conflict that offset any potential benefits. There are also the inevitable questions about who should and shouldn't play God in addressing environmental challenges through geoengineering in a global environment and whether these actions could lead to a false sense of security that might dampen more immediate efforts. For decades, this has been more of a conceptual argument; today, it might be our only option, given our lack of progress in CO_2 emissions reduction and other challenges. Of course, the last thing humanity needs is another Frankenstein path we don't understand in geoengineering. Still, it's probably too late not to utilize some reverse-engineering methods to improve the planet's health. We have little choice, but we need to be careful about it if we wish to achieve these sustainability goals by 2050.

We live in a material world, and whether we like to acknowledge it or not, all of us in advanced societies are materialists, myself included. If someone calls you a materialist, it doesn't necessarily mean you are shallow and overly focused on things. But it's an unconscious and uncomfortable trait when we think about it, and it has become inextricably tied to our lives. There is no going back to simpler times; we can take much of the clutter out of our lives and consume less, but we have reached a point where there's no going back. The same is true for the environment. There can be a more rational approach to human use of materials, but we cannot return to a time when there were no synthetic materials; instead, we need to develop suitable replacements. Likewise, the planet can never return to a time when humans were like other primates, using materials sparingly and without much sophistication. Humans are limited creatures living in a world of finite resources. To manage them better, we must understand our near Godlike control over the planet. But that won't be easy, and it is impossible if we don't start.

Chapter 9

Human Systems and the Environment

Are humans selfish by nature? In his 1976 book *The Selfish Gene*, Richard Dawkins provides scientific evidence of a gene expression that leads to selfish behavior, the survival of the fittest. From a humanist perspective, 17th-century philosopher Thomas Hobbes agreed, noting that social structures and contracts are required strategies to mitigate humankind's violent and selfish tendencies from taking over against each other. Suppose the selfish nature of gene expression drives humans, and the survival of the fittest pervades over all species. Are tribes and communities a control mechanism to mitigate selfishness or the enablement of it? From one perspective, communities and tribes have led to a common good, while tribalism has revealed the dark side of societies. How can we get more of the good and less of the bad? This question is more important than ever as we face the existential challenges related to climate change.

Human tribalism has been a critical element of social evolution. Humans have socially evolved in groups, creating systems to meet basic needs that build trust, dependencies, and cooperation. Trust and dependence were built through tribes when surviving against the elements and other tribes and species, which led to innovations such as language development. The Neolithic era led to the establishment of permanent communities as a result of agriculture, leading to larger group structures centered around food availability and survival against nature, with culture and customs extending beyond basic human needs. Initial forms of religion were developed to keep social order and explain the unexplainable.

An extensive study of human religions that began in 2011 gathered data and artifacts from all the world's regions and found that human societies or systems were formed before religion, which was then used as a unifying factor within a social construct to survive in the world.[1] The study found that belief in the supernatural and the spiritual world is universal and has a meaning greater than stories and explanations for making sense of natural events. Religion became an essential human system for communities to understand and survive these events.

There is an argument that religion is less necessary in the modern world because we understand more of the mysteries of the universe through the scientific method, not tradition. However, according to many studies, an essential attribute of religion and its origins was developing a moral system to regulate the worst impulses of individuals and make communities more resilient to disasters.[2] In the face of an overwhelming world made so large by globalization, religions remain relevant in satisflying the human need to belong to something, particularly among those who do not find much comfort in being called a so-called global citizen. Scientists such as Dawkins scoff at this notion, seeing humans on a biological continuum with bacteria, relatively insignificant. In *The Selfish Gene*, Dawkins notes that "we [humans] are survival machines—robot vehicles blindly programmed to preserve the selfish molecules known as genes. This is a truth that still fills me with astonishment."[3] Because we are, in his words, "survival machines," Dawkins dismisses the role of human systems and an understanding of the social evolution of communities as an escape mechanism from natural evolution. To Dawkins, creating human-built environments is a protection response driven by biological evolution, but could it be vice versa? Yet social evolution has kept us together through communities, and today it is needed to save us from the limitations of our biology. If Dawkins is correct and we are driven solely by biological evolution, then we are doomed. However, if we change the nature of social evolution, something not possible with biological evolution, we can reform our human systems and survive.

Understanding human systems and social evolution makes clear why religion remains popular worldwide despite proof that the ancient texts of monotheism are stories more than anything else. Approximately

84 percent of the world's population identifies with a specific religious group, most of them based on principles established several hundred or even thousands of years ago.[4] Outside of Western Europe and North America, participation in organized religion is increasing in the 21st century. Researchers Pippa Norris and Ronald Inglehart found that religion becomes less central in the lives of individuals when they become less vulnerable to death, disease, and other threats.[5] Other studies have shown that humans have an inherent need to belong and possess an identity as a member of a tribe, regardless of life circumstances.[6] These human systems are the creation of our will to survive, evolve, and belong. As the world feels larger in the face of global conflict and natural disasters from climate change, human systems become more critical, not less. A breakdown of community-based human systems, or tribes, through a so-called world order pulls people back in; the goal must be to do so without the risk of tribalism and danger to the environment. Humans need to be stewards of the planet through communities because the concept of the global community isn't possible.

Nationalism is a version of tribalism that is also on the rise. It is a diluted form of a community-based system, as defined in this book, falling somewhere between it and a global system. Some scholars have associated an increase in nationalism with the COVID-19 pandemic of 2020 and 2021, but evidence shows that as globalization has grown, so has nationalism in response. Nationalist sentiment is growing worldwide, including in the United States, China, India, Russia, and the EU. Global campaigns to unite on climate change, COVID-19, and even the Russian invasion of Ukraine can be viewed as a threat to one's community when it seems too big and outside of one's self-interest. The subgroup of Americans focusing on "Making America Great Again" is responding to a group of "others," or the prioritization of a global agenda over their defined sense of community. Before the environmental topics of CO_2 emissions and plastic waste can be addressed, the root cause of the problems within our communities that spin out of control into nationalism and other variants of tribalism must be acknowledged and understood. In addressing the global challenge of the environment at the community level, we go to the level where problems have always been solved and resolved, starting when we lived in caves.

Attacking the emotions driving these community subgroups, rather than attempting to understand the root causes of their behaviors and norms, does more harm than good. Rationalism and empiricism, while important, are of limited benefit in understanding human systems. For example, my own experiences are based on having been born and raised in Baltimore, which was decimated by the deindustrialization of the 1970s and 1980s. This community has never recovered from the impacts of the global supply chain. Yet because I escaped a dead-end blue-collar path, went to college, and am a white-collar worker, my experiences and beliefs shifted dramatically from those of others who never escaped the city's decimation. Those who feel victimized by globalization aren't necessarily disagreeing with global supply chains and sustainability programs as much as expressing their disdain for its impact on them. When Dawkins mocks those who believe in religion or elitists scoff at the anger of the working classes, they don't understand the human system dynamics. Our human systems need equal intellectual intelligence and emotional intelligence to solve this planetary crisis.

Who better to make this case than perhaps one of the greatest philosophers of all time, David Hume, who warned us that reason must depend on passions rather than vice versa? Despite the value of empiricism and science in solving problems, he believed human emotions and feelings, not logical conclusions, drive behavior. Calling someone a "science denier" because they put their self-interest first or fear losing their sense of community demonstrates a lack of understanding and appreciation of human passions. Increasing interest in and support of religion, nationalism, and other so-called irrational elements of human systems isn't necessarily the result of ignorance or racist ideology, although that can be the case. For some, clutching onto traditions and beliefs in an uncertain world with accelerating globalization and impending environmental doom makes sense. It isn't necessarily irrational superstition, as it is sometimes called. Going to a mosque, synagogue, or church can be as much or more about being a part of a community as it is believing stories written thousands of years ago to explain the nature of the world.

Interestingly, in some regions of the world where organized monotheistic religion is on the decline, there is a growth in paganism, an adherence to animistic beliefs, and a connection to inanimate objects.[7] In those cultures, religion is a matter of connecting to nature, the very claim made to us by scientists and environmentalists! To create change in our human systems, we must possess the very best intellect of the leading scientists and the passion and resolve of those environmentalists and humanists who care for the planet and its people.

Speaking to us from the 18th century, Hume noted that "it is not contrary to reason to prefer the destruction of the whole world to the scratching of my finger," emphasizing the higher priority of the personal self, or even someone in your community, relative to global matters.[8] When climate change policymakers ignore the realities of the individual and communities for the whole world's sake, as Hume called it, they ignore wise teachings about human nature passed down from centuries ago. How good is an understanding within the environmental movement of the relative poverty that exists today, which I estimate to be 70 to 80 percent or more of the entire population? Is the right strategy to propose reincarnation as a virus to control the surplus population, as Prince Philip noted, or that of Mahatma Gandhi, who is quoted as saying, "There is enough on this planet for everyone's needs but not everyone's greed"?[9] One of the most often cited economists of all time, Thomas Malthus, classified the poor as a "surplus population" that needed to be reconciled by God. In contrast, the 18th-century economist Adam Smith saw poverty as a social problem perpetuated by poor economic practices. Different perspectives lead to a critical question for us to answer related to a 21st-century sustainability strategy: are poverty, overpopulation, and environmental damage problems of biology to be addressed by nature through the survival of the fittest, or are they social ones that must be fixed to save the environment? In this book, I make a case for the latter in solving the environmental crisis not through CO_2 emission readings but through human prosperity. I think this position is based on sound evidence.

The Math of Human Systems

Scientists and environmentalists may wish to live in a global world, but there is no evidence that this has been possible throughout human history. From a historical perspective, human systems are most effective when the goals and values of institutions can enable good cooperation and trust within the aggregate population. Unfortunately, there is no evidence that trust and collaboration can happen to address global challenges such as pandemics and climate change. Instead, evidence shows that national human systems are struggling due to a lack of the shared identities, goals, and results that lead to the required level of cooperation and trust. Moreover, there is strong evidence that today's large institutions are not up to solving the most significant problems, given the inherent nature of humans as a species.

The United Nations concept is the most well-known example of globalization. It is a worldwide body representing over 99 percent of the world's population. Almost every nation in the world is a part of the UN; Taiwan and Kosovo are excluded because of political disputes, and Vatican City has a permanent observer status due to its nature as a religious state. This degree of a global organization is unprecedented in history, and it provides some coordination and strategy on critical matters such as the COVID-19 pandemic and climate change. Each UN member is a sovereign nation, however, and the UN has no authority to enforce conceptually binding resolutions. According to a recent Pew Research Center study, the world's view of the UN is primarily positive, with approximately 65 percent favorable, but this is essentially a function of its limited role in people's lives.[10] Views from the Global South are more favorable toward the UN than those from the Global North, which makes sense because the former views of the UN as a protectorate body in response to the balance of power owned by those in the latter. Nations like the United States often view the UN as antithetical to their national, individual, and community best interests. The UN serves a critical yet limited role in the purpose of human systems worldwide.

The next most extensive human system isn't global, per se, but is a virtual international community of 2.2 billion. More than 25 percent of the world's population identify as Christian, but the belief system lacks the structure to have global influence. Catholicism is the largest sect of Christianity, with almost 1.4 billion adherents in total. While the faith is centralized through a hierarchy headed by the Vatican and the pope, there are limits to its impact on individual followers. During the Middle Ages, the church had sovereign control over lands related to political, economic, and ethical rules, but today, it has such no such authority. The church isn't a discrete institution, per se, but consists of numerous distinct subsystems centered only around the teachings of Jesus Christ, as interpreted and defined by each denominaton. There are no jurisdictions, and in some cases, there are sects at literal and virtual war with each other, which is evidence of a lack of commonality and trust. Similarly, the religion of Islam has 1.9 billion followers, or nearly 25 percent of the world's population, broken into different, often disputing sects. Organized religions hold sway over a large percentage of the population, providing a moral code for their followers to uphold but possessing little sovereignty over their virtual communities.

Some large, populous countries exert a significant influence on the entirety of the world's population but are focused on national identity and struggling to do so. For example, the most populous countries in the world are India and China, both over a billion people. Can a nation of 1.3 billion have a human system of government that enables shared identity and values to allow sufficient trust and cooperation? Achieving shared values and cooperation across such a large population is difficult despite strict governmental measures. Presently, the Chinese government struggles with multiple domestic issues beyond climate change, such as a shrinking population, the Uyghur population, stubborn income inequality, declining economic growth and opportunity, global geopolitics and supply chains, and lingering problems from the COVID-19 pandemic. In this model, if trust and cooperation aren't possible among the general population through shared identity and goals, it will be induced in the population through state-run means.

The current most populous nation in the world, India, faces enormous internal challenges, including a chaotic government structure and capabilities, large and growing income inequality, and regional and cultural conflicts. Prime Minister Narendra Modi has often commented that he proudly represents the world's largest democracy. Still, researchers challenge this notion; the Swedish political research group V-Dem called India an "electoral autocracy."[11] One perspective is that Modi has a relatively high level of public support, at least with the majority Hindu population, not due to the effectiveness of his government and its policies but rather because of the large and growing nationalist sentiment among Indians.[12] The same seems true in the next two largest nations/unions, the European Union, with 447 million inhabitants, and the United States, with 340 million citizens. As is noted in this chapter, there are fractured communities in each, given a lack of shared values and trust. There are growing flashpoints of nationalism and tribalism that have led to the disillusionment of governments to lead their people. The EU is considered by some to be the stalwart of global elites intending to supersede national sovereignty through policy, the greatest enemy to the nation-state in the world.[13] In Russia, authoritarian president Vladimir Putin seeks to restore the glory of the Soviet Union by reconnecting the former satellite nations, which are balking in favor of their national identities. The weaknesses of these large public institutions are a distraction from the need to focus on addressing our existential environmental threats.

Much of the remainder of the world, including the nations of the developing Global South, has some combination of ineffective state government systems, insurmountable challenges in a global economy, national identities established through inappropriate geopolitics as a result of colonialism and other misadventures, and other challenges. Other institutions or human systems of note are the private enterprises of the financial system, the networked global and national systems, and the multinational corporations that exist as accountable not to citizens but to their shareholders, as designed. In many cases, what's best for consumers and shareholders isn't in the best interests of the workers and citizens of the world.

Maybe it surprises you that you're reading a book about 21st-century sustainability that is focused this extensively on human societal matters. But we must understand that although perfecting solar panels and high-tech food production is the more straightforward solution, concentrating on solving our environmental challenges predominantly through technology has been a thirty-five-year mistake. Follow the course of history and watch how tribes that turned into larger institutions solved problems but could go no further. In the 21st century, the balance of responsibility for fixing the issues of the environment has been in these large institutions. Now it is time to acknowledge the limitations of these institutions.

Solution #1: A 21st-Century Model of Globalization

The 19th-century philosopher Karl Marx noted that "the tradition of all dead generations weighs like a nightmare on the brains of the living."[14] This quote is taken from his 1851 book *The Eighteenth Brumaire of Louis Bonaparte*, which discussed the coup d'état by Napoleon III, a mediocre, bureaucratic politician who paled in comparison to his uncle, the original Napoleon Bonaparte, the French emperor often remembered for the many improvements he made after the French Revolution. Progress moves forward and then backward, as we see in the rapid advancement in globalization that today is driving us in the wrong direction, from both a geopolitical and an environmental standpoint. Another example are the efforts made by Abraham Lincoln to keep the union together that were relatively abandoned by his successor, Andrew Johnson. As Marx noted, traditions and institutions of the past can be a weight on society that is both good and bad. Can our traditions of the past, a sense of community, lead to a model of community systems networked with other communities lead to a sustainable 21st-century model of globalization? To do this, we must take what's best of local problem-solving without resorting to tribalism; we need to take what's best from the past generations and leave the rest behind, if possible.

Advances in technology might be the key to taking the best and leaving the worst within a community-based system. Sustainable 21st-century globalization might also be the best solution for networking communities worldwide, rather than relying on large public and private institutions that rely on the unproven reliance on large-scale human consensus and alignment. The definition of a virtual model of human systems in the 21st century can take advantage of technology to enable physical communities to link virtually worldwide through enabling technologies such as the internet, blockchain, and big data. We can create a new platform to solve problems differently and disparately. Individuals and communities can network more effectively, rather than relying solely on large public and private institutions through conventional means.

A 21st-century definition of human systems or even a 21st-century definition of globalization can be enabled through technology that enables individuals and communities to drive change in the economy and the environment. For change to be effected by individuals and consumers, technology must become less capital intensive and more open sourced, rather than overly proprietary and controlled by multinational corporations. There is a need for improvements to education so everyone in society can become a part of a digital solution and to avoid a worldwide digital divide. As a result, this model is a paradigm shift in the definition of globalization as it is understood today, which is based on economies of scale and large institutions. Noted 20th-century globalist Thomas Friedman, author of the globalization bible *The World Is Flat*, wrote a 2021 *New York Times* op-ed asking the question of whether the United States, China, and Russia would work together if one of the nations were invaded from outer space, suggesting that climate change is a similar threat.[15] Four months after Friedman's article appeared, Russia launched its "special operations" against Ukraine, leading to a proxy war between Russia and the West, with the world on the brink of a nuclear war. Despite Friedman's desperate pleas over nearly thirty years for the world to come together—first through a global supply chain and more recently through a conceptual scenario of aliens from outer space—it was and is never to be. We need virtual globalization to pervade where the old model has failed in facing our shared challenges.

Through virtualization of the network and commonality within the community, individuals can build alignment based on shared values rather than focusing on areas of differences. Rather than reforming failed human systems, this solution is an alternative to an institutionally driven model. Instead of continuing to work through the UN, the World Economic Forum, nation-states, and multinational corporations, accountability for their own best interests is placed in the hands of communities and individuals, as Smith designed. As Marx noted, nothing is stopping us from this new 21st-century model of globalization other than the traditions of dead generations weighing upon us.

This solution requires a strategic plan that is an internet model of a state, nation, or continent in terms of how to focus globalization and environmentalism within communities, as will be discussed in the remainder of this book, and how those communities are linked. It is an alternative model to the behemoth, overwhelming model of a monolithic global community.

Solution #2: 21st-Century Sustainable Finance

The Committee to Save the Planet is a triumvirate of some of the world's most critical financial leaders: Mark Carney, former governor of the Bank of England; Larry Fink, CEO of BlackRock, the world's largest investment firm; and Jamie Dimon, CEO of JPMorgan Chase, America's largest bank.[16] If you have faith in the capability of large institutions to solve the climate crisis, this is a veritable dream team: three influential and passionate global leaders committed to addressing this existential crisis through financial markets. Yet when we separate their good intentions from the human systems they lead, the contradictions become apparent, as they are bound by their fiduciary responsibility to their shareholders and not public policy. In 2022, a German consumer group sued Deutsche Bank's asset management group DWS for alleged greenwashing in misrepresenting a fund's sustainability accomplishments.[17] These sustainability-induced ESG programs may be well intended, but they address the symptoms rather than the root cause of the problem of corporate finance. Can traditional financial markets change their nature to address the world's sustainability challenges? For example, the world's largest energy producer, Saudi Arabia, showed profits of $110 billion in 2021 from its state-owned oil company Saudi Aramco and pledged $187 billion to the green economy, but without a plan or a timeline to stop drilling for oil.[18] In Smith's model of capitalism, the financial resources should flow to the area of most significant opportunity for both the public and private good. Public and private institutions are quick to tout their sizable investments into the so-called green economy and ESG funds, which sounds impressive in news releases—but what is the complete business case calculus that includes the cost, benefits, and comparison to the alternatives? Their words have spoken louder than their actions.

Today's financial model remains stuck in the 20th century, which enables large, entrenched public and private institutions to control the flow of investment. Large government institutions such as the United Nations, NGOs such as Greenpeace, large financial houses such as JPMorgan Chase and BlackRock, state-owned oil and gas companies such as Saudi Aramco and Gazprom, private oil and gas companies like ExxonMobil and Shell, and state-run utility companies around the world are notable examples. According to a 2022 McKinsey report, the green economy requires a $9.2 trillion investment annually, $3.5 trillion more than today.[19] To provide context around this investment, it would be almost 10 percent of the world's GDP, half of its corporate profits, and 25 percent of tax revenue. Furthermore, while investments in the green economy are growing, so are global oil and gas revenues, projected at $5 trillion in 2022.[20] This approach to financial analysis is the wrong way of looking at the problem; rather than a high-level institutional analysis viewing environmental policy as a behemoth global transaction, how about community-based opportunities flowing to areas of the most significant opportunity, such as the continent of Africa? Today's outdated 20th-century economic model doesn't flow to the areas of greatest opportunity in a manner described by Smith. Large institutions will continue to hold sway over financial markets, but we need suitable investment alternatives to them, especially given that it is highly unlikely the world will invest 10 percent of its economy in sustainable activities during a time of so much immediate need.

Yet there is some evidence of change. As a silver lining to Russia's invasion of Ukraine, the green energy sector is gaining momentum to address the instability of the oil and gas markets, which will be a benefit to the environment. Increased investment in renewable energy may lead to power capacity growing by 2,400 gigawatts between 2022 and 2027, equal to the total power capacity of China.[21] Just one region in Europe, the North Sea, has the potential to power millions of households with green energy and still have enough left over to manufacture green hydrogen for transportation. Today, it is becoming increasingly likely that green investments and technologies will become more viable and profitable. Impediments in today's financial markets are not necessarily due to investment in these fields—such as solar and wind energy, sustainable food, and new materials—but are the result of the financial system's effectiveness. According to an FTSE Russell report on sustainable investments, the market capitalization of green investments is now at 7.2 percent of the total, more than double what it was just five years ago, but it is highly concentrated in technology investments and the superpower economies.[22] A concentration of investment in technology and economic superpowers, such as the United States, China, and EU, indicates a bias toward assuring financial return on investment for its stakeholders rather than solving the problem of climate change. For transformation to happen in sustainability, the financial system must be liberated from these large institutions to enable a more efficient flow of capital to opportunities across the planet.

According to a 2022 International Energy Agency (IEA) report, governments spent $1.2 trillion in clean energy investment support between April and October 2022, more than double the financial commitment since the 2007–2008 financial crisis.[23] Looking at the IEA report, much of the funding is in subsidies for existing businesses and government spending, such as fleet improvements, consumer incentives, and other incremental improvements. Disruptive innovations might be more difficult to cost justify under current market methodologies, so we need new strategies to change the structure of financial markets to meet our 2050 objectives.

A conventional definition of "sustainable finance" is focused on environmental, social, and governance principles in investment decisions rather than individual and community-based investing in disruptive innovation. While such terms are endearing and promising, there is little evidence of "boots on the ground" where they are needed the most. In her 2022 book *The Value of a Whale*, Adrienne Buller expresses the absurdity of "green capitalism," such as carbon taxes and financial models posed by entities such as the International Monetary Fund (IMF) that place a price tag on whales.[24] In Buller's model, the current market-based system of economics must not be distracted by carbon and species accounting and instead show returns on investment in solving problems that generate profit within financial markets. Buller is skeptical about the capabilities of large public and private institutions to drive change but doesn't offer satisfactory solutions other than grassroots protests. I agree with her on her problem assessment, but merely identifying a problem isn't sufficient; we need new capital market structures to get there.

By definition, large institutions must tend to the current state to meet the needs of their existing consumers and investors, and this must be acknowledged and accepted. However, if the goal is to change the plans within the system, the current state must be disrupted by an outside agent. The flow of financial resources must become more peer-to-peer to direct incentives to innovators outside the mainstream, which is the key to disruptive innovation. For every sizable social entrepreneur who can invest in ESG-related projects, I'm projecting an immense potential for funding flows from direct peer-to-peer transactions. The goal must be to divert the trillions of dollars needed for disruptive innovation to achieve sustainable solutions from government agencies and multinational corporations that might be well intended but are focused on conflicting and incremental answers. We need frictionless capital and more funds invested into the projects themselves.

While it's an unproven concept at the present state that must mature, a decentralized model of finance (DeFi) might be the best application of Solution #1. DeFi is a term used to describe the application of blockchain in a peer-to-peer model. Rather than a dilution of funding by large governmental bureaucracies, these funds could go directly to innovators for maximum benefit. There are estimates today that the green economy is funded up to $50 trillion, but what is the balance, effectiveness, and yield of these investments related to the overall market to meet our 2050 environmental goals? Of course, individual investors and governments will continue to have the ability to fund using the conventional institutional model. Still, the research shows these traditional investment models are not sufficient in scope to address the problem. An alternative model for direct, peer-to-peer investment using newer, more efficient technologies, such as blockchain, is necessary to address these challenges. We need disruptive, not incremental, innovation, and for it to happen, more money needs to be invested more effectively as soon as possible.

As an additional note, the topic of DeFi and other alternative modes of finance disparate from the conventional model has come under great scrutiny arising from the FTX/Sam Bankman-Fried debacle of 2022. There is no doubt that the decentralized nature of this approach to finance requires greater regulation, control, and understanding. Still, it would be a mistake to throw out the baby with the bathwater, given its potential to reform today's financial system, which is equally disastrous. Over the past decades, there has been no evidence that the current capital market model is equipped and incentvized to address the 21st-century sustainability challenges. Instead, today's financial system model mirrors the mercantilism model Smith discussed in the 18th century. If we are to embrace Smith's principles, as many in the finance world say they do, we need to employ this solution rather than a continuation of today's highly concentrated institutional model of finance and economics. Through more peer-to-peer and peer-to-cause solutions, there is greater opportunity for the disruptive innovation that is required for this transformation!

SOLUTION #3: THE NEW UN (UNITED COMMUNITIES): OPEN-SOURCE ENABLER

Solution #1 is a new platform of globalization that is defined as a technologically virtual system, a self-organizing networked system in which individuals and communities participate. While this platform doesn't displace the existing UN, it must lead to significant reform of this mid-20th-century relic. The general perspective of the UN is that while it is an imperfect organization, it is a necessary institution. As noted by the second UN secretary-general, Dag Hammarskjöld, the UN wasn't created to lead humanity from heaven but to prevent it from hell. This quip made sense in the middle of the 20th century as the world recovered from a global war, but it is time to expect more from this organization. After accounting for inflation, the cost of the UN is forty times higher than it was in the 1950s. The UN includes seventeen specialized agencies, fourteen funds, and a secretariat with seventeen departments that employ approximately forty-one thousand people.[25] The UN has done good, but more is necessary given these expenditures. Rather than disbanding it, which would require something to take its place, the UN could be the facilitator of a new globalization platform focused on benefits to communities and individuals rather than strictly focused on the national level. The old UN model, which made sense after World War II, is not viable today.

This new vision of the UN will be challenging given that the organization is called the "United Nations," not the "United Communities," and given that organizations exist as human systems, meaning it is improbable that they would agree to limit their roles. The goal is for the UN to gradually reform itself toward this model as an approach to becoming more individual and community centered and to use technology to improve collaboration and cost-effectiveness. Because the UN has higher approval ratings than most countries, there might be a reluctance to reengineer it in such a drastic manner. Still, while the body enjoys overall favorability from the general population, it is viewed as less effective by the major parties funding its efforts, including the United States, and its accomplishments aren't commensurate to its costs. If the goal is for the UN to remain relevant into the 21st century and beyond, it must reform itself to perform better on global matters such as environmental policy, poverty, and economic development. Therefore, a transition from more of an institutional, bureaucratic body to more of a virtual, networking system could salvage its role in the future.

In 2015, the UN adopted a blueprint for its Sustainable Development Goals, seventeen goals critical to healthy societies and the planet (see chapter 2 for more details). By any assessment, these goals by 2030 are not only off the mark but, to many, out of reach. Furthermore, the impact of the COVID-19 pandemic and the Russian invasion of Ukraine is likely leading to a lack of progress toward sustainability goals and a backtracking on progress already made. The UN lacks any authority in accomplishing this mission other than creating strategic alignment, as it did with its 2030 SDGs. However, some incremental progress toward the UN enables an open-source platform to achieve these goals. A 2021 UN resolution, "open-source technologies for sustainable development," called for a collaborative platform to perform practical tools and technologies toward its SDG goals.[26] This is a good concept, but as usual, it is limited in meaning and execution. Rather than passing a watered-down resolution, the UN should redefine itself as an enabler of an open-source platform to achieve the goals it develops in strategies it has no authority to implement. Using the massive amounts of funding it receives and its worldwide credibility, the UN could be the game changer of peer-to-peer, not through the dream of a global community, as it has tried for nearly eighty years, but rather as the platform provider enabling innovators to do so. The UN could create an open-source platform for innovators worldwide to collaborate on SDGs, and even facilitate funding for these solutions.

The good news is that an open-source enabler developed by the UN would make it a viable leader in the innovation platform worldwide. However, to some extent, this could be seen as a threat to national sovereignty or, even worse, as a conspiracy to create a global world order controlled by the elites. However, a platform of networked communities would improve national and international collaboration without impinging on sovereignty. It shouldn't be viewed as a threat by anyone other than those benefiting from the highly concentrated power and authority of large public and private institutions preventing progress toward these goals. If such a project could not be facilitated by the UN, then perhaps it could be created by a social entrepreneurial investor. Investors such as Bill Gates have donated billions of dollars to innovators toward important global causes that have made a difference, but none have funded an innovation platform, per se. Funding innovation has led to benefits but has not necessarily accelerated the needed innovations discussed in this book. An open-source platform could accelerate the innovation cycle, allowing investors to fund these projects.

There are questions regarding open-source versus private innovation related to intellectual property, a hallmark of the modern system of capitalism. To balance the needs of the economy and society, we need both, using transparency, security, and efficacy. As an intergovernmental body, and consistent with private multinational corporations, the UN could play an enormous role in creating an open-source model to complement and compete with closed private and public systems, not replace them.

It would also offer a more robust form of innovation to today's largely incremental model, given the need for governments and corporations to balance the present and the future. Universities and think tanks could also play a role in this model, enabling basic and applied research to flourish. While I agree with critics of current capital and research models' ability to solve these lofty challenges, we have to work within the present system to develop the future; developing models that work in theory isn't sufficient.

Solution #4: Africa Strategy

The African continent has been cursed for centuries while other regions have prospered. Geographically, Africa has wide diversity in climatic zones, leading to more variation and decentralization in culture and socioeconomics, which has prevented the development of larger, more populous nations such as those in Asia, North America, and Europe. Slavery was a complex problem facing Africa that stemmed from large outside empires from Europe and Asia, which led to the continent becoming an unfortunate element of a global supply chain. For example, the Europeans developed a slave trade with West Africa that shipped people to the Americas, traded slaves for the natural resources grown there, and then traded the natural resources from the Americas for manufactured goods from Europe. This trading system led to the depopulation of the African continent, which impacted economic growth. A depopulation of the African continent through slavery led to the weakening of the continent, enabling colonialization that lasted through the 20th century. As a result of its history and geography, Africa is the least developed region of the world and yet is paying one of the largest prices from climate change.

Africa is an example of a failed human system that isn't just the responsibility of those who live there but also of the rest of the world. Today, Africa has forty-eight sovereign nations, or approximately 25 percent of the world's total—the same as Asia—but only 16.7 percent of the world's population, compared to nearly 60 percent in Asia. An analogy of the problem of Africa is consistent with that of the global climate crisis; there has been progress toward improving the welfare of the African people, yet it is an incremental improvement versus the transformation that is needed, such as what has happened in other regions of the world like China. Given its history, some have called for reparations or investment to improve its economy, with the former as an expression of anger and the latter as a call for hope. To meet 2050 sustainability goals, Africa must be front and center as a foundation for this strategy. Foreign direct investment must be made, without strings attached, through redefined financial markets, as noted in Solution #3. Richer nations need to invest in Africa as a matter that benefits all nations; the continent can become a proof-of-concept innovation center for the new sustainable economy!

According to a 2022 report produced by the IEA, Africa must double its electricity generation capacity to have universal electricity access by 2030. Today, not only is electricity consumption in Africa exponentially lower than in the developed world, the capita energy use for an African is less than the electricity used to power your refrigerator.[27] Addressing this problem will require significant amounts of public and private capital, which have been promised at levels of $100 billion a year for more than a decade but haven't yet materialized. African economies are growing primarily through raw material commodities and built infrastructure, and Africa is in need of a healthier manufacturing and service-based economy. Much has been discussed and promised at the numerous conferences concerning economic development and sustainability, and it is too often framed in colonial methods rather than equal, collaborative partnerships. Africa is the second-largest landmass in the world yet perhaps the most important. Improvement in human systems in and supporting Africa must be addressed. The world needs to have a different perspective and strategy toward this continent. The methods discussed in this book related to energy, food, water, and materials must begin in and have a central focus on Africa.

If the world is serious about addressing the challenges of climate change, ESG, peer-to-peer, and governmental funds must flow to Africa—not just because of the dangerous nature of the climate change challenges to the continent but primarily given the wide-open opportunities for commencing energy, food, water, and material strategies without the human system resistance that is inevitable elsewhere. Investing a more significant percentage of environmental sustainability funding in Africa would lead to a positive ripple effect worldwide.

Solution #5: Nonsovereign Nature Strategy

Our human systems have failed the areas of the world that aren't sovereign, such as the oceans, the Arctic, Antarctica, and the atmosphere. Furthermore, humanity continues to grow its footprint into parts of the world that used to be shielded from the built environment, leading to an imbalance that is bad for both the economy and the environment. The UN approved its Convention on the Law of the Sea in 1982, but it has not been ratified by all nations, including the United States, that govern the oceans more as a resource than for protection. Today, countries continue to fight over ocean sovereignty, particularly the Arctic Ocean, which has competing claims from six countries. Russia claims more than 70 percent of it, primarily for its mineral rights.[28] Over the last forty years, legitimate actions have not been undertaken to develop rights and responsibilities over the ocean, which covers more than 70 percent of the planet. And in contrast to parts of the earth that are contested from a resource perspective, nobody is responsible for the atmosphere and the open ocean—nor does any nation care to take responsibility for it. The jurisdiction of the planet is driven primarily by resource potential rather than environmental stewardship.

What is the solution to the problem of areas of the planet that aren't under sovereign protection and responsibility? And what should be our strategy in protecting natural ecosystems outside of the built environment for the entire planet's health? In the next chapter, I will discuss a bifurcation strategy between the built and natural environments to improve the health of the planet. This strategy is antithetical to the biophilia hypothesis formed by biologist E. O. Wilson, who formulated the concept in 1984. I agree with Wilson's hypothesis that humans have an affinity for being connected with nature, but it's impractical to take that to mean eight billion people can live within or even adjacent to it. We can visit nature to be among it without living in it; our best method for protecting nature is to organize cities and suburbs separate from nature to enable both to regenerate and flourish.

This last strategy is provocative but can be a game changer in how we view our industrial, built environment and the planet in its natural state. One should not be considered more important than the other, but rather we should have a balance between the two. With eight billion people living on the planet, we can never return to Thoreau's Walden Pond, but we should be able to enjoy it to build the connection Wilson discussed. We need to pull the impoverished into the cities for the betterment of themselves and the environment. This is our best strategy for a 21st-century balance between economics and the environment.

Humans have succeeded throughout our history through systems of social development. Yet we remain limited in our capabilities to work together in the dream of a so-called global community. Rather than developing concepts that consider conventional globalization as a possibility related to climate change and other manufactured damage to the environment, we need to work within our limitations and innovations. The previous three chapters provided specific solutions to specific environmental problems. Yet science, technology, and supply chain innovations are the easy part of meeting the goals by 2050; getting eight billion people across 195 nations in various states of prosperity on board is the hard part. The first step is recognizing that a new paradigm must be put in place that addresses these human limitations.

In the next chapter, I will lay out a plan moving forward from today to the year 2050 for developing and implementing this 21st-century sustainability strategy. Science must be the inventor and supply chains the innovators, but human systems must be addressed to drive both. Humans have advanced science to the extent that we can send multiple spacecraft to Mars, 56 million miles away, to observe the planet. Yet our human systems are incapable of feeding the world population when food production is 150 percent of what is needed. And while we continue to push farther away in exploring other planets, nearly two-thirds of our planet hasn't been discovered—and too much of what we have discovered has been destroyed. Something is wrong with this picture, and once we understand, it makes sense that we are facing a fate relative to the planet's health and sustainability. Education, research, and cooperation to improve these human systems, more so than technological advances, are key to solving the problem.

Developing a roadmap to confront these challenges over the next twenty-five years is doable if we start with the foundations of structured problem-solving and focus on the root causes of these problems. Economic growth, globalization, industrialization, and carbon emissions are symptoms. A lack of understanding of the relationship between relative poverty and environmental damage, the limitations of human communities, and the illusion of globalization are the root causes. Innovation and technologies can be solutions if they are focused on the problems of outdated human systems. There must be a change in priorities, with more responsibility given to individuals and communities than to large institutions. The 21st-century strategy plan I provide in chapter 10 is different in both method and solution, and we need a paradigm shift in thinking if we expect transformational change. "If you change the way you look at things, the things you look at change."[29] Let this be our mantra for moving forward!

Chapter 10

A Blueprint to 2050

We must develop a detailed, actionable, achievable plan for sustainability to meet the 2050 target. An existing blueprint comes from the International Energy Association (IEA), a robust 224-page roadmap for achieving carbon neutrality by 2050.[1] IEA members include thirty nations, mainly in the Global North, and eight associated countries, including Brazil, India, and China, but not Russia, Saudi Arabia, and other large energy producers. The IEA is a recognized authority on global energy matters; still, it doesn't have any jurisdictional authority even within the 20 percent of nations that are members or affiliates and certainly not within the 80 percent of countries not included. The 2021 report, developed by leading researchers and authorities in critical fields and peer-reviewed by more than eighty public and private world organizations, is impressive and worth reading.

I agree with much in the IEA report, such as the quadrupling of wind and solar projects, an eighteen-fold increase in electric vehicles by 2030, and a 4 percent reduction of energy intensity for economic growth year over year. From one perspective, this blueprint is too conservative and not ambitious enough, and furthermore, it is already behind schedule. For example, the report called for a ban on new oil, gas, and unabated coal capital investment by 2021, an unrealistic expectation, and it didn't happen. The IEA report is almost entirely focused on existing or developing technologies for 95 percent of the solution and human will and behavior for only 5 percent. In this report, there is an overemphasis on technology and not enough emphasis on human factors.

In my blueprint, at least 50 percent of the solution is focused on human systems and behavior. In the IEA study, "human behavior" was generally defined as end-consumer activities, such as taking the bus to work rather than driving and not taking long-haul flights, solely incremental improvements. Yet it did not consider more difficult challenges, such as curbing subsidies, laws, and regulations prohibiting innovation, and changing consumer, business, science, and government behaviors. Too often, environmental proposals emphasize the low-hanging fruit, such as not to take a bus or using a plastic straw, while at the same time, an environmental law that enables water waste and fossil fuel subsidies is not a consideration. To achieve dramatic improvements, we must move forward on bold yet practical solutions. The actual "inconvenient truth" is our unwillingness to accept human limitations and develop strategies to solve problems through public and private institutions. Proposing feel-good solutions such as CSR and ESG reporting applications may seem easier—still, it is these uncomfortable challenges, or the existing shortcomings, which I will discuss in this chapter. Technology is not always the answer, and as we learned from Dr. Frankenstein, it can sometimes be the problem. It's not artificial intelligence or some energy transformation that will lead to environmental change; it's us—we have to understand ourselves differently to bring sustainability to the planet.

The People Plan to 2050

Perhaps the most uncomfortable of all topics is a strategy for the planet's built and natural environments. E. O. Wilson's biophilia hypothesis suggests that humans are inherently associated with the natural world, but what this means moving forward must change. The planet cannot withstand Wilson's connection to the planet of a population of eight billion. A 21st-century symbiotic relationship must move more humans away from the natural environment into cities so the population rate of those living in relative (and extreme) poverty stabilizes through infrastructure improvements and leveraging the future strategies mentioned in this chapter. Rather than economic growth through continuously tearing down natural environments necessary to protect the planet, we must separate geography between built and natural. As a result, humans will have more opportunities to visit nature than live within it. A bifurcation of constructed and natural may be provocative, but the math and science suggest it's our best option given the current and future data trends. Separating the natural and built environments is the foundation for my strategy. We need to enable the natural environment to heal, such as increasing carbon sinks.

The evidence is irrefutable that governments cannot drive the environmental change as required; instead, they must become facilitators of the 2050 goal. Yet the IEA report holds governments accountable for "emission reduction targets" and "net zero pledges" despite the lack of evidence; governments are evaluated based on whether they have made a "nationally determined contribution" (NDC) toward emission reduction, have made net zero pledges, and have long-term strategies and laws. Practically all nations have made NDCs, such as being a signatory of the Paris Agreement, but only ten have defined net zero as a legal obligation in a policy document; eight of those ten have followed through in making it legal, but there's no evidence that any of the eight have made progress toward the 2050 goal. No evidence exists that national governments can lead the effort to achieve these 2050 goals.

A better role for governments is to incentivize and facilitate policy to fuel innovation and innovators, with a particular focus on individuals and communities. It is also clear from the data that large multinational corporations cannot drive this change either, given their fiduciary responsibilities to their shareholders, who are focused on short-term financial gains. The relationships between large public and private institutions favor the sectors that need to be reconsidered, such as fossil fuels, utility companies, and farmers/ranchers. The IEA report is rather vague in its mention of "green jobs" and does not state how the economy's focus should shift from a 20th-century industrial to a 21st-century green revolution. My plan is to establish a new market model of community-based systems that will lead to glocal employment to benefit the community. How else better to engage individuals and local communities into sustainability than to tie their economic prosperity directly to it? Today's model of globalization has been harmful to both the environment and workers; a community-based energy/food supply chain model can achieve greater economic and environmental balance.

We must also take on the complex topics related to public and private institutions. For example, today, trillions of dollars in fossil fuel reserves belong to nations and private companies, and hundreds of billions of dollars in publicly funded subsidies steer energy strategy. In modern society, governments and companies have property rights that might be difficult to resolve. Ignoring these matters because they are difficult impedes the 2050 roadmap. While there are no easy answers to these modern-day quandaries, tackling them on head-on will lead to better options than ignoring them. Governments must lead conversations that change long-standing laws, subsidies, property rights, and other laws that have stymied environmental innovation; otherwise, all of these so-called climate pledges are pointless.

Lastly is the issue of population and poverty challenges. Significant population declines are forecast in many parts of the world, including China, which used to be the world's most populous nation. While many environmentalists lament over an increase in the worldwide population, perhaps up to 10 billion, demographical forecasts show a stabilizing and declining population. Additionally, there has been a rise in poverty levels after COVID-19, and trends seem to indicate that it is continuing on the wrong trajectory. Therefore, by 2025, there needs to be an economic and sustainability strategy to redefine economic activity in a way to significantly reduce poverty levels. By 2030, the goal should be reducing extreme poverty from 9 percent to 5 percent, significant poverty from 46 percent to 40 percent, and moderate poverty from 82 percent to 70 percent. By 2040, the goal should be to eliminate extreme poverty and reduce significant poverty to 30 percent and moderate poverty to 60 percent. Finally, by 2050, considerable poverty should be reduced to 20 percent and moderate poverty to 50 percent. If any aspect of globalization is to be encouraged, it should exist relating to poverty elimination.

Eliminating extreme poverty and substantially reducing significant poverty is the best solution for stabilizing the environment. Lowering population growth rates in the Global North and moving more of the Global South into community-based systems centered around cities will lead to a healthy decline in population and poverty rates, also leading to greater protection of the environment. The world's population living in cities could increase from 56 percent today to 80 percent by 2050. Concentrating a more significant portion of the world's population in cities would allow large geographic regions to become natural again, away from the built environment, freeing space for plants and animals to regenerate. In addition, focusing people within robust community-based systems in cities and suburbs would offer a better balance between humans and biological systems.

Building networked, community-based systems concentrated in cities can organize populations to achieve reductions in poverty. The goal would be to focus these communities in areas of the fastest population growth—namely, on the continents of Africa and Asia—starting in 2030, after a blueprint strategy is developed from 2025 to 2030. First, research must be conducted to establish the logistics and funding for these regions. Then financing must be established, and these communities should be built and networked in clusters of centralized-decentralized systems, much like the internet. It is undoubtedly an ambitious project to focus on redefining economic activity to address poverty in this manner, and many details need to be addressed. It will require collaboration through enabling technologies that will raise suspicion across populations and be questioned. However, a human systems strategy of centralized decentralization, taking advantage of improvements in process and technology, and concentrating populations in city centers has benefited economic growth and the environment. Biologists, zoologists, and environmentalists have complained of humans crowding out ecosystems and threatening to create the sixth mass extinction. If we concentrate humans and human systems into cities, we need better institutions serving individuals and communities. There are many details to work through between 2025 and 2030, but, rather than scientists planning the colonization of Mars and other distractions, this is our best hope.

Strategies and Design: 2025

Addressing these fundamental challenges related to people and their systems will be crucial to the solutions I am proposing. In general, the time frame for answers to these problems is to develop strategies and design work by 2025, commence the implementation of these designs and strategies by 2030, and implement these strategies by 2050. In the remainder of this chapter, I will define the process to do so and the goals for each within these stated time frames. However, given that it's already 2024, there is a need to transition as quickly as possible in this direction!

Strategy #1: A 21st-Century Reinvention of the UN

The first strategy is a redefined, 21st-century focus on the environment and sustainability of the United Nations. Currently, the UN has three main areas of focus: the Intergovernmental Panel on Climate Change, its Sustainable Development Goals, and agreements through its annual Conference of the Parties. I have provided evidence that the design of large, global, intergovernmental organizations such as the UN is ineffective, given the nature of human systems and our intrinsic successes in solving problems through communities. The United Nations is the most trusted organization in the world and has a global platform; with that in mind, the goal of Strategy #1 is to develop global platforms through the UN to enable environmental and climate innovation. The platforms to be developed will be an open-source model for information/data sharing used for material design and management and a network design that will link the community-based systems. Given the broad level of trust and support the UN enjoys worldwide, it can be a good fit for the UN to be the provider of this platform if it chooses to transform itself. Think of the UN as a 21st-century high-tech platform to educate, inform, incentivize, and lead change rather than just talking about it. If not, these platforms and networks can be developed and led by another global organization, perhaps newly developed, that can earn the trust and respect of nations and their people. This first strategy to implement by 2025 is to create a strategy that will lay the foundation for future initiatives, as noted in this chapter. The goal of this strategy is to build a global community bridge to the best extent possible.

Strategy #2: Free-Market Economics via the World Trade Organization
Beyond reducing poverty and bifurcating the built environment from the natural world, an essential solution must be to liberalize free-market economics in the way Adam Smith designed them. The World Trade Organization (WTO) is a relatively effective global organization that works with its 164 member countries for alignment and agreement. To be a member of the WTO, a nation must agree to specific domestic and trade policies that include tariffs, property rights, fairness, and transparency. The WTO cannot require any government to follow free-market principles; instead, it establishes frameworks. Strategies #5, #6, and #7, which will be addressed later in this chapter, can succeed only within a liberal free-market system as defined by Smith. For the WTO to uphold its role in enabling fair and effective world trade and functioning national economies, it needs to consider the importance of focusing on the environment and the economy, which it hasn't considered at this point. For all the discussion of free-market economics, it's time for the talk and the action to be consistent. If poverty is the greatest threat to achieving environmental stability, a case I have made in this book, the WTO needs to take leadership over liberalizing markets to accomplish this goal.

Strategy #3: Alternative Platforms to Traditional Finance Such as DeFi
In 1944, near the end of World War II, a conference was held in Bretton Woods, New Hampshire, to establish international finance rules of order to prevent the chaos that was a contributing factor to the Great Depression and World War II. Among other agreements of this conference was the creation of the International Monetary Fund to regulate and manage payments and currencies, and the World Bank to provide loans and grants to lower-income nations worldwide. By virtually all accounts, these efforts have led to some improvements in economic stability and equity but have at the same time exacerbated issues of poverty, inequity, and climate change. Nations in the Global South have complained of the so-called debt trap of receiving funding when their economies are in crisis and having to agree to harsh and untenable terms that are difficult, if not impossible, to overcome. Likewise, much funding has been funneled through the World Bank related to climate and food security matters,

leading to unfair conditions biased toward significant institutional government interests. A study conducted between 1970 and 2004 indicates that less-developed nations receive more World Bank projects when they occupy a rotating seat on the UN Security Council.[2] These large institutional organizations are responsible for addressing public policy matters such as poverty, climate change, and the environment and have less-than-stellar performance records given the amount of money invested in projects. More recently, China has also been accused of creating these debt traps in its grand Belt and Road Initiative, raising concern regarding conventional funding for strategic projects. Today's financial markets have been ineffective in deploying capital to the areas most in need, which is what Smith requires in his definition of capitalism.

A decentralized finance strategy is needed as an alternative to conventional markets to drive capital flows consistently in balancing the public and private interests in a liberalized market economy. A DeFi strategy is decentralized, allowing individuals and communities to invest in projects without the use of centralized entities such as private banks, countries, and intergovernmental organizations. For the most part, DeFi has been associated with cryptocurrencies. However, given the recent challenges related to the destabilization of these currencies and their systems, there may be a movement away from a DeFi strategy for public and entrepreneurship funding. This is a mistake; greater controls should be put in place, and the goal must be to attract more individual and community investors and entrepreneurs whether funded through cryptocurrencies or traditional funds. In addition, the goal is to divert more current state funding that flows from governments to intergovernmental institutions, such as the World Bank, toward more entrepreneurial processes, such as directly to social entrepreneurs within or for a specific community. These 20th-century funding processes have been ineffective in addressing societal and environmental challenges, and the goal is to use technology to flow more capital toward entrepreneurship and achieve sustainability investments at a lower cost. This strategy is a networked approach to finance between individuals and communities worldwide.

Strategy #4: A Global Materials Genome Database and Network Tied to Sustainability Targets

In chapter 8, I discussed the concept of the Materials Genome Initiative to sequence and document each material element, raw material, alloy, composite, and compound. Various nations have undertaken their own initiatives to improve the time to market and effectiveness of material science within their industry. However, for a 21st-century sustainability model, these systems must become networked, as the material flow of raw materials to composites is worldwide. In addition, these databases must become integrated with other tools, such as sustainability targets for recycling and reuse, which I will discuss in terms of a bill of materials process and database in Strategy #5 and for use in better managing products as services in Strategy #6.

Imagine the possibilities of an integrated network that houses a materials genome design of all the world's materials, alloys, composites, and compounds ever discovered, sequenced, invented, and published—not on one site, and with some materials proprietary and others in an open-source model. Elements and raw materials shouldn't be a problem to map and publish because they cannot be intellectual property, and other materials are in the public domain. As a starting point, a global body such as the UN can develop and house a database starting with these materials in an open-source manner. (If not the UN, it could be any entity creating the database as an open-source model.) The eventual goal is for all natural and manufactured materials to be mapped into this database, with some as public domain and others protected based on their intellectual private or public property rights. Without undertaking this effort, it cannot be possible to create, produce, manage, reuse, and dispose of materials in a manner to support a 21st-century sustainability strategy on a finite planet with a population surging past eight billion people. A solution must be enacted that provides transparency about the planet's materials, natural and human made, while also respecting intellectual property.

This strategy, to be developed by 2025, must determine how to optimize a public open-source model while at the same time protecting private intellectual property. Blockchain is a technology that allows centralized-decentralized transactions, meaning rights can be established based on rules. Using this technology, it seems possible to enable individuals and communities to access materials based on these rights within a centralized-decentralized network. For example, if a material is owned by a private enterprise or even a government, an individual entrepreneur can be granted or pay for access to develop a solution. Tied to the environment, by documenting the characteristics of a material, technologies such as artificial intelligence can determine how it can be recycled/reused, and if it cannot, new materials to replace it can be created. In this model, materials that are harmful to the planet, such as polyvinyl chloride, also known as "poison plastic," can be better controlled within a closed loop or replaced with new material. In addition, data can be gathered and analyzed relative to materials for a practical approach to sustainability. Without a global materials genome database and process strategy, we cannot achieve our 2050 goals.

Strategy #5: A Bill of Materials Process and Database

Building on Strategy #4, a bill of materials (BOM) is necessary as a list of ingredients within a product. You have seen lists of ingredients on food and drink labels and medication. However, there are no requirements for BOMs beyond these consumables, which leaves out hundreds of thousands of finished products, if not more. Look around and you'll see a lot of different products that aren't mapped; only the manufacturer of that product understands some (but not all) of its materials. It makes sense to document food, drinks, and medications because we ingest them into our bodies, but these aren't the only materials with the potential to impact our health and that of the environment. If we wish to develop a 21st-century sustainability strategy, we need to understand everything our society and the planet are ingesting, so to speak, to control the flow of these materials. Otherwise, there's no hope that we won't destroy more natural ecosystems and run out of materials. For example, according to a study published in the journal *Environmental Science and Technology*, researchers found that rainwater across the planet has concentrations of toxic chemicals such as per- and polyfluoroalkyl substances (PFAS) that exceed safety guidelines.[3] It's difficult to comprehend how big a problem these chemicals and other synthetic materials are in our lives and the natural environment, and the problem is only getting worse. Establishing a BOM for every product designed and produced provides the transparency required for human and natural environmental health.

Given the authority the WTO has over the world's economy and its supply chains, it makes sense for it to be responsible for establishing the standards, protocols, database, and disclosure process. Some private companies may push back against a BOM concept, suggesting it will put their companies in legal jeopardy, which is ironic given the consumer liability associated with so many chemicals unabated in the world. A strategy developed through the WTO could take into consideration both public and private interests and rights, but material transparency is necessary for a 21st-century sustainability strategy. Products, supply chains, and their impact on public and environmental safety are a significant concern in the 21st century, and the rise in cancers and endocrine disruption, among others, is a topic that needs to be addressed quickly in a transparent manner.

Strategy #6: A 21st-Century Definition of "Product" and "Service"

Next, we need to address the definition of "product" and "service" in the 21st century. This can be accomplished through four subsolutions: closed-loop systems, product fixability, product sharing, and products as a service (PaaS). The first, closed-loop systems, should be a strategy deployed for any materials that are used only once or a few times or have a low value after their end of use. For example, single-use plastics and everything else you throw into your recycling bin need to work within a closed-loop system. Through Strategies #4 and #5, companies that develop single-use plastic and other throwaway materials should be tasked with either creating a closed-loop system for their products or redesigning them to make them more sustainable. In the former case, materials that aren't easily recyclable in other waste streams can be closed to prevent environmental damage or waste, and in the latter, new product designs will give the material economic value for secondary purposes. The second substrategy is product fixability, which relates to a consumer being able to fix an existing product rather than buying something new. Products that fit into this category include appliances and technology devices, such as smartphones, computers, printers, and other more durable products. While these products can be repaired, the market is designed in a way that makes it suboptimal to do so. Washing machines, dryers, printers, and other items used to be built to be fixed, but today, it is often not viable to do so. The concept of planned obsolescence can lead to higher revenue for companies, but in the 21st century, there needs to be a better balance between what's best for the economy and what's good for the environment. The WTO should develop a strategy that enables this balance. Increasing sales while doing damage to the environment should be viewed as unethical behavior.

The last two substrategies speak to less ownership of products: product sharing and products as a service (PaaS). Product sharing involves things like renting a chainsaw or a pickup truck at a local Home Depot, car sharing memberships, and community tool libraries. With this strategy, the goal should be for all products, not just large capital investments, such as vans and power equipment. Smartphone applications can be introduced to make it more convenient to share products rather than to buy them, especially for infrequently used items. The last substrategy is PaaS, which includes such things as using Uber rather than buying a car, renting a Vrbo rather than purchasing a second home, or even hiring a lawn service rather than buying a lawn mower. These models are a form of de-materialization, or eliminating the things in your life. An economy and environment built on overconsumption and obsolescence will ruin us if we keep at this pace. Change must inevitably happen soon. In its revised role, the WTO should emphasize these reforms to balance the economy and the environment better.

Strategy #7: Legal and Legislative Reforms

In this book, I have identified various legislative and legal hurdles that are detrimental to developing a 21st-century sustainability strategy. First, in G20 countries, significant subsidies are provided to the fossil fuel energy and agriculture/ranching sectors that hinder innovation in competition such as renewable energy and alternative food production processes. And secondly, in many countries, private property laws, such as those in the United States, incentivize an agricultural operation to waste water so as not to lose the allocation related to its land ownership. For this strategy, the WTO should take a leadership role in introducing legislative and legal policy reform efforts to better balance the requirements of a healthy global economy and the environment. The WTO website makes references to Adam Smith and his book *The Wealth of Nations*, identifying the economist and his writings as foundational to the organization. The Adam Smith Institute, likely the highest authority on the 18th-century economist's theory, published an article on its website titled "We Agree Entirely, Let's Abolish Subsidies," making the argument that the $1.3 trillion spent annually is driving the destruction of wildlife and heating the planet.[4] Therefore, if the WTO truly lives by the words of Smith, it should undertake a significant reform effort to reduce these expenditures worldwide. As is too often the case, subsidies are being used to not do something rather than incentivizing burgeoning industries such as renewable energy and vertical farming as it is for existing, entrenched sectors such as big oil and big agriculture. This is undoubtedly a complex topic to undertake in a world full of nations that have difficulty cooperating, but it must be addressed.

The WTO's involvement in reducing and eliminating subsidies to large, entrenched industries that are bad for the environment should be a slam dunk; however, the perspective is less clear regarding the intersection of private property rights and the collective nature related to the environment. Today's so-called free-market capitalists suggest that Smith fully supported individual rights with no qualification, but that wasn't necessarily the case. Smith had no illusions that markets were perfect. When he railed against large institutions, it was regarding the monarchy and mercantilists, but he was not necessarily saying that individuals aren't accountable to society. Instead, he suggested that individual liberties and actions should coincide with what's best for the community. Land ownership and property rights are crucial elements of Western capitalism and should be protected, but they can also enable what's best for the collective in the process. It's time we redefine these terms, which haven't been revised for centuries.

Finally, improvements must be made in how the government works relative to permitting and applications for innovation, technologies, and incentives for improvement. There are too many stories of new technologies and innovations being slowed down by government bureaucracy at a time when we can least afford to move slowly. We need to act as if our collective future depends on our actions—because it does.

Strategy #8: Public Utility Reforms

The 20th-century model for public utilities has outlived its useful life here in the third decade of the 21st century. Electricity, gas, and water have shifted in the United States from centralized government models to more of a privatized market, but they remain monopolies due to infrastructure and regulation through managing commissions. These systems are often classified as "natural monopolies" because the infrastructure is most efficiently funded through single-source capital investments. For energy, utility companies are responsible for manufacturing energy (power generation) and distributing it (electrical grid) as one integrated system. Water utilities operate in a similar way; the collection of water in dams and reservoirs is integrated into the distribution of water within a specific municipality. In this model, utility regulators, often through a utility commission, develop agreements to balance the financial health of the companies and the needs of the citizens in a centralized manner.

In the United States, the Federal Energy Regulatory Commission (FERC) regulates electricity and natural gas, seeking to balance the needs of the current electrical power generation and grid with the future of distributed energy resources (DERs). A federal government directive in 2020, FERC 2222, enabled small DERs to operate alongside the traditional monopolistic public utilities, a game changer in moving toward renewable energy sources and the community-based model discussed in this book. Federal agencies such as FERC and the Department of Energy focus on renewable sources, distributed energy, grid reliability, interconnectivity, and everything necessary for the 21st century; however, implementing these reforms often encounters opposition from special interests of the large oil and gas companies and large utility companies such as Pacific Gas and Electric, a $16 billion company in California. These large private companies, operating as monopolies and managed through state and other municipal commissions, are a labyrinth of complexity and self-interest that impedes progress toward a more efficient, community-self-reliant, and environmentally friendly system, and they should also be more interconnected across states and nations for greater resiliency. If renewable energy is important, governments need to level the playing field and stop enabling powerful special interests.

Although public utility rates in the United States and around the world continue to rise, that money is not going toward enabling a centralized-decentralized DER system that is necessary to meet economic and sustainability objectives. As monopolies, public utilities have little incentive to move in this direction of decentralization, and federal directives such as FERC 2222 are not leading to sufficient progress relative to these goals. Public utility reform needs to segment power generation, which should be emphasized as centralized and decentralized from power distribution via the grid, which should be used as the network for this system. Because public utility firms can only make their financials work as a centralized monopoly, they wish to control and regulate the network of power generation and distribution rather than share control with local communities and individuals. The restriction of decentralized power generation is a significant roadblock to advances in renewable energy. In many communities, even if you place solar panels on your home, you are required to subsidize the costs of the centralized system. In the new model, if a residence or business can generate sufficient power for its needs, it shouldn't be under any control but allowed to purchase energy from the main grid at market rates. It should be allowed to trade energy to a central grid for what's best for the economy and the environment. The public utility could act as the hub in a hub-and-spoke network, but not control (disincentivize) localized and distributed energy resources. Energy reforms should be implemented to enable this model for the future, and large public and private institutions shouldn't get in the way. The role of today's energy monopolies is the greatest barrier to renewable energy.

Solar and wind energy should be freed from monopolistic utility systems. If this issue isn't addressed soon through legislative reform, the technologies for power generation and storage will be sufficient to achieve carbon neutrality, but the human systems will not be able to support this transformation. As with Strategy #7, we need a strategy to reform these existing monopolies that provides the necessary progress to make this possible. For the water companies, a similar problem needs to be addressed. These public utilities are focused on only one source of water and less focused on water reclamation, rainwater utilization, and other nonpotable sources that will be necessary for the present and the future.

Strategy #9: Energy Generation and Grid Strategy

Tied to Strategy #8, there should be a worldwide (as much as possible) networked model design for the electrical grid that is consistent with that of the internet, using DERs. Other than in nations that restrict access to their citizens, data flows on the internet in a centralized-decentralized manner. Today, there is some connection of the power grids that span boundaries, but it is limited, for various reasons. In the past, energy generation and distribution made sense to be centralized, given the capital intensity and requirements for fossil fuel as its feedstock. Today, both variables have changed, and it makes sense for individuals and communities to take the lead, with the role of centralization as some loose form of networking to share resources. There will never be one global network for energy. However, as with the internet, there is an opportunity to develop a network cluster of DERs tied to 21st-century storage devices and networked across public utility power grids for distribution. This design needs to be the future of electricity generation and distribution.

This 21st-century design must be optimized using new technological innovations. For example, the energy grid in the United States is 33 percent efficient, meaning two-thirds of the energy generated is wasted through poor distribution and lack of storage. Through a smart grid design networked worldwide as much as possible with renewable forms of feedstock (solar and wind), more energy can be used to support growth while improving the impact on the environment. An internet-like global design (as much as possible) based on AI and renewable feedstocks would be a game changer for the economy and the environment, and an international body such as the UN should begin the strategy and design for its development. While the internet isn't global, it is an effective communication tool used around the world, so it could be the basis of this design for the future of energy. Strategies should be undertaken to build this design that could unleash a new model for energy in the 21st century!

Imagine the future of the electrical grid: a networked, centralized-decentralized design powered by renewable and nuclear energy and managed through AI and network optimization tools. Tied to this new design of electricity generation is the creation of green hydrogen through surplus power for portable fuel. Because hydrogen has the highest energy mass of any fuel, it can be used for portable transportation that requires more power than is optimal through electrical batteries, such as an airplane. Today, there are significant and understandable concerns regarding a transition to renewable energy, such as the intermittency of sunshine and wind. These concerns have led to some even suggesting the world will always need fossil fuels as a significant element of an energy strategy. These concerns do not take into consideration that the root-cause problem isn't the limitations of the energy feedstocks, but rather the limitations related to an end-to-end energy system; if this is addressed, we will be able to achieve 100 percent renewable energy by 2050, or sooner!

Strategy #10: Agricultural Strategy and Design

Thomas Malthus was wrong. As food production increases, population growth declines and economic prosperity can grow. The problem isn't population growth but rather the distribution of the food grown, which is 150 percent of what is needed to feed the world. And even though food production per hectare has improved exponentially, large land areas such as the Amazon rainforest are now being cultivated for the wrong reasons. Existing farmland is becoming less productive and more reliant on harmful chemicals, given today's industrial production processes.

In addition, agriculture is primarily driven by large-scale global supply chains that perpetuate significant food waste levels and an uneven flow of food through commercial import/export relationships.

Similar to Strategy #9 related to energy, a strategy to improve food production and use should focus on better distribution methods, better purposing of stock, and the use of technology to monitor and manage the process for a network of food communities to large industrial producers in a hub-and-spoke model. A 21st-century strategy for farming shouldn't be to displace industrial agriculture but to use it as the hub attached to numerous community spokes. This will reduce import and export transactions and farm subsidies, all of which hurt innovation and discourage local self-reliance. Developing a hub-and-spoke model of food production enables each community to use enabling technologies, such as vertical farming, to grow more food locally that is also sustainable. Today's local farms are often worse for the environment than industrial farms, but a new model of vertical farming can change this conventional model. Large industrial farms can remain in place but must be responsible for economic and environmental improvements to enable innovation without being subsidized.

A hub-and-spoke, centralized-decentralized, community-based model should commence in areas of greatest need, such as Africa and the small island nations. Food security should be tied to local self-reliance; currently, some regions of the world are wholly reliant on other areas. Using 21st-century agricultural practices will allow localization because a vertical farm, for example, isn't dependent on weather conditions and other variables that create disadvantages in the global food supply chain. A hub-and-spoke model will enable a better flow of food while reducing the high level of food waste today, which is close to 20 percent of total production. A healthier food production system requires less water, energy, and chemicals; moving in this new direction will improve the strategy of these models concurrently. Furthermore, today's inequitable food production and distribution system is a foundational problem that impacts global security; enabling a balance between large economies of scale and local self-reliance will lead to greater political stability and less fighting between countries over natural resources. Creating a centralized-decentralized, community-based system is a win for communities and individuals.

Today's food market is a disconnect between a private enterprise that leads to improvements in production and efficiency and a lack of public good strategy that leaves hundreds of millions in starvation while half the food we produce is misallocated. An example of this disconnect is using agriculture to create transportation fuel when hundreds of millions worldwide are starving. Rather than legitimate market principles, subsidies enable agricultural production for the energy markets (ethanol) and meat production. Using agricultural grains to feed animals to produce meat is necessary for today's economies-of-scale food supply chains but isn't sustainable with increased consumption. Likewise, government regulation in the United States requires a certain percentage of ethanol in gasoline as a political motive while corn for use as food is needed worldwide. By propping up prices artificially rather than making improvements in efficiency and innovation, poorer nations of the world must pay higher costs for food; in some places, families spend over 50 percent of their income to eat. It's an example of how poverty and its negative environmental impact go hand in hand. A rational model for the economics and supply chain for food should be created through a 21st-century strategy.

Strategy #11: A 21st-Century Meat and Dairy Strategy

The challenge of feeding humanity has changed since the days of Malthus and Smith in the 18th century. In those days, there were maybe a billion people in the world, with roughly 90 percent living in poverty. Most people ate whatever crops grew in the area, and fewer people ate meat grown on a farm, where it was sustainably grazed in a meadow. Today, there are eight billion people in the world, with a smaller percentage in extreme poverty and a growing share of the population eating industrial meat grown by feeding animals agriculture that could be used to provide food for people. As such, eating meat is one of the worst things someone can do relative to the health of the environment. According to the UN Food and Agriculture Organization, aggregate global meat consumption increased by almost 60 percent between 1990 and 2009.[5] The planet cannot withstand the existing global supply chain process of meat, much less any further growth, as will undoubtedly happen.

Today's raising, slaughtering, processing, and distribution of meat are not only bad for the environment but also bad for us. Because there is such a high demand for meat on such a large scale, animals must be tightly packed together in factory farms, not grazing in the idyllic fields of the past. Fish and shellfish demand is growing so fast that we are depleting our oceans and creating aquaculture farms that pack in the fish and create nasty wastewater that flows into natural water bodies. Antibiotics are used extensively in managing these food systems to prevent widespread sickness. As a result, these drugs are becoming less effective in fighting infections in humans. The close quartering of animals and the integration of these farms into human populations have also made us susceptible to animal-to-human viral infections, such as COVID-19. When we add together the impact on the environment as well as our individual health risk, it is clear that meat is a hazardous element that must become a primary focus for a 21st-century sustainability strategy.

The consumption of meat will always be critical for humans, both in our need for protein and for its cultural importance. This means we need to become creative in defining what a "meat supply chain" must be in the 21st century. Those concerned about the future of meat grown from cells, or cultivated meat, should tour an existing cattle ranch, chicken-raising facility, slaughterhouse, and aquaculture farm. What you will see more closely mirrors other large industrial manufacturing facilities than the pastoral settings of the past. Chicken-raising facilities are indoors, where the animals are tightly packed in, receive antibiotics in their food, and live in controlled conditions, leading to widespread avian influenza outbreaks. Many of us are also concerned about the ethical questions about the treatment of animals raised for food. A new strategy where animal cells are grown to create food addresses most of the challenges associated with our present-day food system, but it raises other concerns. The primary concern of a cultivated meat approach to a 21st-century food system is that our current meat is more than just the cells of animals. Animals become meat through a developmental growth process, even in industrial conditions; the process of raising cattle and chicken that will become filets and steaks has been expedited, but it's still the process. By 2025, humans need to develop a strategy for changing the definition of meat to describe animal cells successfully and safely grown in a lab. Cultured cells can be produced in a hub-and-spoke model as well, leading to the optimization of food production between global supply chains and local communities. Growing meat from cells is an uncomfortable topic, much like others that I am recommending, such as bifurcating the natural and built environments. But if we are really serious about an environmental strategy, these are the topics that must be addressed.

Strategy #12: Water Strategy

The need for a water strategy falls lower on the list than energy, even though it is also often classified as a public utility. This is because water use in today's global systems is highly concentrated for industrial use in energy, agriculture, and meat production, so the problem will be addressed to a large degree simply by creating more effective and efficient energy, agriculture, and meat systems. In addition, water utilization and yield through traditional farming practices are highly inefficient. Finally, in many countries, there aren't laws to properly manage water use for the public good, allowing landowners to do as they wish. Many of these challenges will be addressed with the implementation of earlier strategies related to water rights and laws and more effective energy and food production and distribution.

The days of the water cycle we learned in school—evaporation, condensation, cloud formation, and rain—are over. However, simply having enough water from the skies deposited onto fields for crops or into lakes and other freshwater sources is not enough and is unsafe; given the proliferation of chemicals, nobody in the world should be drinking untreated rainwater. Underground aquifers are another primary source of water, and they are being depleted at alarming rates. And yet water on the planet flows in a closed-loop system, meaning the earth isn't running out of water; we are misusing it. With more than 70 percent of the planet covered in water, there's plenty of water for all life if we develop a 21st-century water strategy that is much different from today's finite and ineffective approach. There's also plenty of nonpotable, nonsalinated water to convert.

The first element of an effective strategy is to address water usage in scope areas such as energy, agriculture, and meat production—not just in terms of lowering use through laws that protect the public good, not using water to process fossil fuels, and addressing waste in farming and the out-of-control meat production system, but through better measurement and management processes to improve yields, and a higher rate of recycling and reusing water. This also should include an education program for the population, as some who propose the reclamation of wastewater without understanding that this water source is much cleaner and safer than unprocessed rainwater. Potable water is a precious commodity and must be managed as such. We no longer live in a world where we can expect the water cycle to be sufficient alongside groundwater; we must be more efficient with our potable water and develop an effective strategy to manage it.

Secondly, we must develop a strategy for nonpotable water to include other sources since only a tiny sliver of the planet's water is potable. This contains effective processing techniques for brackish and salinated water, of which there is a near-infinite bounty on the earth. Regarding seawater, the present-day challenge of desalination is the amount of energy required for the process and the need for a strategy to deal with the residual saline. Through large-scale production of renewable energy through solar and wind, this process to develop freshwater will increasingly become viable. Like the energy problem, a water strategy must take into account both the acquisition and treatment (generation) of clean water and its distribution as two different issues. Many nations, such as Pakistan, are experiencing water extremes of intermittent droughts and flooding. A networked, community-based, hub-and-spoke system can enable local supplies to ebb and flow through rivers, lakes, dams, and other solutions. There will be limits to the ability of a global community to manage water. Still, through a healthier system with an effective strategy, it seems likely that fewer geopolitical conflicts will arise if supply is better addressed to demand. Water shortages and supplies are some of the most significant flashpoints in the world leading to many disputes;

a 21st-century strategy for water will not only improve economies and human welfare but also lead to improvements to the environment and less geopolitical conflict.

Strategy #13: A Chemical/Synthetic Strategy

Our world is full of synthetic chemicals, including those being rained on us from the sky. Finding a place on earth free from chemicals and synthetic materials—including at the top of Mount Everest and in the deepest trench in the sea—is impossible. Traces of plastics and other synthetic materials are everywhere, including in the bodies of wildlife and humans. At birth, a baby born today has mystery chemicals in its body. More than sixty years ago, Rachel Carson wrote the book *Silent Spring* to warn us of this problem, and since then, our use of synthetic materials has only grown in scale and scope. The world we live in today is very different from the world of the 1960s, and we need a strategy for the 21st century to consider this. In discussions regarding the environment, there is often a focus on CO_2 emissions, yet the challenges we face are much more significant and complex.

Strategies #4 and #5 will be critical to addressing the synthetic and chemical challenge: an MGI that documents and defines all elements and chemicals on the planet, and a BOM database and tracking system process to determine their uses. Today, there are hundreds of thousands of industrial chemicals in the world (nobody knows for sure), and few are understood and managed from a material flow standpoint of production, use, and post-use. When it comes to our food, the ingredients are tracked because they enter our bodies. Yet today there are so-called mystery chemicals in our bodies. The human body today contains 109 synthetic chemicals, 55 of which had never been reported before and 42 are mystery chemicals from unknown environmental sources.[6] For example, a recent study in 2022 found that freshwater fish have levels of PFAS that are 278 times higher than commercially farmed fish.[7]

A strategy needs to be developed by 2025 for managing these chemicals and synthetic materials in a global supply chain for the health of humans, other species, and the planet. Materials must be redesigned if they are unsafe in the environment and our bodies or managed in a closed-loop system to keep the planet—and us—safe from harm. Private enterprises can no longer design and distribute synthetic materials that harm the public good. The production and use of industrial chemicals is out of control when water from the sky and fish from a river can no longer be consumed.

Strategy #14: E-Material Strategy

A dilemma we face in the 21st century is our increased use of electronics, or e-materials. From a certain point of view, these materials are beneficial for the environment, such as a smartphone that replaces fifteen or twenty different products and batteries that have the potential to be powered by renewable energy rather than fossil fuels. On the other hand, electronic materials have grown exponentially in both scale and scope, with the demand for raw materials such as lithium swelling to levels that require us to find alternative sources. Given that many end products require numerous components and raw materials, often in trace levels, recycling and reusing these materials is complex; a recycling and reuse strategy has not kept pace with the increase in production and use of these products. As such, our landfills are littered with e-waste, which contains valuable materials that can no longer be used and have the potential to cause damage to the environment that isn't yet understood.

An e-materials strategy must be developed to address this growing 21st-century problem. Products must be designed using the MGI and include a BOM to improve their recyclability and reuse. This is one of the most crucial strategies to develop from a timing standpoint, given the exponential growth in production and use of these materials that will continue into the future.

Strategy #15: Metals and Minerals

Metals and minerals should have higher recycling rates than the other materials mentioned in this chapter. Metals and their alloys are relatively easy to recycle and reuse, given their melting points and processes to re-alloy and form. Aluminum, for example, is 100 percent recyclable and reusable, yet in nations such as the United States, given its supply chains, aluminum cans are recycled at a rate of only approximately 50 percent.

Therefore, the problem isn't in the product's material composition, but rather in the supply chain system itself. Due to these issues, trillions of dollars of valuable aluminum is buried in landfills. A similar case can be made for concrete, which is mainly used in large-scale, practical, industrial uses and is 100 percent recyclable but is often recycled and reused at a much lower rate. Again, many of these materials are used in industry and on a large scale, but programs aren't in place to reuse them at the 100 percent level that is consistent with what is possible.

Strategies #4 and #5 will support higher recycling and reuse rates for metals and minerals. However, beyond these strategies, a more effective system must be developed specifically for each element, composite, and alloy to ensure we use existing materials before digging up more from the planet. For example, recycled aluminum is infinitely more sustainable than virgin aluminum because this element is rare in nature, meaning that producing virgin aluminum requires a lot of energy to transform the bauxite rock into alumina and then into aluminum. Burying valuable material in a landfill while further environmental damage is being done in bauxite mines in Australia and China is an out-of-balance scenario in terms of what's best for the economy versus the environment. Developing a 21st-century strategy for metals and minerals that uses what we've already extracted from the planet rather than wasting it and digging for more is necessary. Strategies #4 and #5 will support the better usage of these important materials, but important organizations by commodity, such as the Aluminum Association and others, must establish strategies for their materials. There can no longer be an acceptance of valuable materials to be landfilled out of convenience.

Strategy #16: A Strategy for the Global South

The following three strategies are tied closely to each other. They relate to the biggest challenge we face in the environment: the relationship between humans and natural ecosystems. Malthus first suggested the impact people experiencing poverty have on the environment in his 1798 work *An Essay on the Principle of Population*, in which he predicts the growth in the human population would exceed the potential for food production, leaving humankind in a hopeless condition. In 1968, biologist Paul Ehrlich associated this strategy with the environment, finding that a high population of people experiencing poverty leads to the ruin of the natural world. Both theorists, and countless others, have made wrong predictions about what would happen, and none have led to solutions. And yet the beat goes on within the environmentalist movement claiming that the problems within impoverished countries have to do with their policies rather than the nature of geopolitics and the global supply chains that are driving these challenges.

Too often, today's global supply chain systems treat the Global South as a cheap manufacturing site and a rich natural resource. Despite a drop in extreme poverty, levels of poverty remain high in these nations, and their economies are primarily dependent on global supply chains rather than self-sufficiency and local self-reliance. Developing a strategy for a community-based system that is networked as a global hub-and-spoke model allows Global South economies to be less reliant on large public institutions and private multinational corporations. For globalization to succeed in the future, it must be in balance within communities that perpetuate local self-reliance for the betterment of the economy and the environment. Supply chains must be linked in a network that seeks balanced consumption and production rather than viewing some parts of the world as producers and others as consumers. In this sense, the Global South is the key to a healthy global economy and environment.

As much as it has become a topic of conversation, insufficient efforts have been developed to create a Global South that is both locally self-reliant and networked within the larger global community. This strategy is necessary for the future, codified as important through investments from the wealthiest countries to make it happen. These investments aren't charity or even reparations for the past, but rather a wise investment in a more peaceful world and healthy environment. Countries such as the United States should not view the rise of the Global South as a threat, but rather as an opportunity for its balance and growth in the future.

Strategy #17: Bifurcation of Built and Natural Areas

This strategy might seem controversial, but it is founded on data from the past century. To an environmentalist, there is no better dream than to live in the country again, as Henry David Thoreau did when he left society and lived on Walden Pond in the mid-19th century. But, of course, the dream must end when we factor in simple math. When Thoreau lived on Walden Pond, the world's population was a little over a billion people, meaning that today, the planet has more than eight times the number of people living on it. Therefore, if we wish to respect and make the natural world healthier than it is today—with a population of eight billion—we need to bifurcate human civilization from much of the natural world, with a large majority of the total population living in cities. By doing this, we can concentrate the human population within urban centers where each individual's carbon footprint would be lower and free up most of the remaining land for nature to regenerate in a healthy manner that would be best for the planet.

Humans could experience the natural world by visiting it rather than living among it that tears down natural ecosystems, ruining them. With larger areas of the planet restored to their natural state, the earth would regenerate, which would assist in creating a significant carbon sink to absorb carbon dioxide, helping the world achieve carbon neutrality while at the same time using fewer hydrocarbons through the efficiencies of concentrated human populations. People living in cities have lower carbon footprints than those in rural or suburban areas, and population growth rates are lower in cities. This strategy has many details to develop, such as existing land ownership, and human freedom to live where they want. How to implement this strategy must begin in the spirit and understanding that it is best for both the world economy and ecosystems to do so, and then we can discuss how to make it happen. A strategy shouldn't be set as some form of a mandate, and there's inevitably a conflict between what's best for an individual versus society. There are social and cultural issues to address, as some cultures are based on more rural geographies. Family structure and association can be lost when living among eight billion people versus residing within your tribe. This certainly would shift how human society develops, and its relationship to the natural world, but few other options make sense. If we continue on our current trajectory, with humanity taking over more of the natural environment, it is not feasible for any sustainability solutions to make a difference by 2050. Determining a strategy to make this viable will be a challenge, but it is entirely necessary for an effective 2050 sustainability strategy.

Strategy #18: A Focus on Nonsovereign Regions of the Planet

Tied to Strategy #17, this is a goal for the health of nonsovereign areas, such as the oceans, rainforests, polar regions, and geographies, shared by more than one nation. Strategy #17 aims to increase natural land areas not tied to human society and industry. This would be a significant paradigm shift in our way of thinking about our role as humans on the planet. For example, by concentrating human populations in cities, less rainforest land would be bulldozed to create communities and farms. Likewise, if agriculture and meat were produced using vertical farms and cultivated meats, fewer large-scale industrial farms and ranches would be needed, allowing more land to be restored to nature. While these natural lands may continue to be part of particular nations, they would become separate or sovereign from human development and industry, and rules on the balance of this would need to be determined. This is a difficult concept to acknowledge given the importance of private land ownership in the world, but it must be addressed.

According to the International Union for Conservation of Nature, 15 percent of the earth's land and 10 percent of its territorial waters are protected by national parks.[8] A 2022 research report published in the journal *Science* notes that 44 percent of the earth's land requires conservation efforts to enable healthy biodiversity and ecosystems.[9] To develop a 2025 strategy, the human system must be designed to understand and define terms of private property ownership rights and public sector/environmental benefits that will ultimately be a sticking point in developing and implementing these changes. Humans must live effectively, efficiently, and equitably in more concentrated regions of the built environment, a shift from when economic growth and opportunity occurred through expansion and sprawl. Creating healthier nonsovereign and sovereign zones for nature makes sense for both the economy and the environment. Still, conflict over what's best for the individual versus what's best for the planet may arise. Furthermore, these nonsovereign parts of the planet need to be better understood, such as the large percentage of the ocean that is unexplored. There is no question that this will be a most challenging issue to resolve, but once again, it is necessary. We need to better understand nonsovereign areas of the planet to develop a successful strategy.

Strategy #19: Supply Chain—Manufacturing

The last two strategic initiatives focus on the global supply chain; the first is manufacturing, and the second is logistics and transportation. Today's version of manufacturing within global supply chains is a relic of the 20th century and a misapplied understanding of Smith's intention in the concept of free-market economics. After World War II, the US economy produced more than half of the manufactured goods worldwide. That figure declined throughout the remainder of the century, with production migrating to Europe and some nations in Asia, such as Japan, South Korea, and Taiwan. At the start of the 21st century, global manufacturing capacity was unleashed, with the many nations in the world becoming involved. As a result, there is currently a glut of manufacturing capacity through a global market driven by cheap labor, effective logistics, and emerging technologies. During the COVID-19 pandemic, the world learned how broken this approach to global manufacturing can be when there are shortages in critical medical supplies, such as personal protective equipment and pharmaceuticals. However, the worldwide manufacturing and supply chain crisis—including the deindustrialization of the Global North and the commodification of labor and the environment in the Global South—began much earlier. Globalization wasn't flat but spiky for societies and awful for the environment.

Today's outdated 20th-century manufacturing is centralized, global, and based on cheap labor, mass production, high waste, and environmental damage. Results are focused on short-term economic gain concentrated in large public and private institutions that are not optimal for individuals and communities. Therefore, a new strategy needs to be developed to create a community-based system that is glocal, balances consumption and production as well as the public and private sectors, and reduces waste. Through a network of these community-based systems,

3D printers can be used to make products within a community, balancing inbound materials and finished products for higher recyclability and reuse. By networking these community-based systems that use 3D printing and are connected to large industrial global manufacturing, an optimization in mass production and customization can be achieved to balance supply and demand, production and consumption, and the economy and the environment. It may be logical to continue to globally mass-produce some products, such as cheap commodities that can be recycled and reused within a global supply chain. However, critical products, such as pharmaceuticals, food, energy, and other essential products, should have some production within communities for their own local self-reliance, supply resiliency, and societal health. When nations, regions, and communities are overly reliant on others, the imposition impacts that society, as has happened (and is happening) in Africa. In bringing local manufacturing through technology that is also balanced with a global industrial capacity, there is an opportunity for private enterprise profit to be better balanced with local employment health for workers. American entrepreneur Peter Thiel famously noted that "competition is for losers," making a statement out loud that some businesspeople endorse.[10] However, such statements do not represent liberal free-market economies and the optimization of supply and demand for community benefit.

Manufacturing grew during the 19th and 20th centuries through consolidation and globalization, but as we have learned through COVID-19, it offers the world diminishing returns in the 21st century. Large private multinational corporations are winning in concentrated and globalized production, seeing competition as a weakness, which is undoubtedly the case from the perspective of their self-interest but not from an overall market perspective, as Smith noted. Suppose free-market capitalism is the optimal solution for the 21st century—and I believe it is. It must be democratized by enabling networked communities through emerging technologies such as 3D printers. Once again, it becomes a challenge of necessity for socioeconomics and the environment.

Strategy #20: Supply Chain—Transportation

In the 19th and 20th centuries, transportation became a critical element of a globalized and industrialized supply chain system leading to today's world economy, which is approximately $100 trillion. On any day of the year, numerous steamships are moving across the ocean, planes are flying in the sky, railcars are heading down the tracks, trucks are on highways and streets, and personal vehicles are moving us around. People, products, raw materials, components, money, and information flow from one point to another, with physical movement that used to take weeks or months happening in hours or days, and the movement of money and knowledge, which used to take days and weeks, often happening in a matter of minutes or seconds. The good news is that the world is more connected than ever before, leading to a greater understanding through trade, travel, collaboration, and the exchange of information. The bad news is that this transportation of people, materials, money, and information is destroying our planet, particularly in physical movements across the sky and ocean that require an enormous carbon footprint. For better or for worse, transportation is the straw that stirs the drink.

Transportation should remain a process of good through exchanges of culture and information for better collaboration and understanding. However, travel needs to be facilitated through renewable methods, such as solar-powered batteries, wind, and even green hydrogen as a replacement for oil and gas. The other important element of transportation is its reduction through community-based systems. These networked, community-based systems have a hub-and-spoke model with globalized manufacturing supply chains and other features. However, it's best for both the economy and the environment if a more significant percentage of production and consumption occurs locally. Today, there are circumstances where this is not the case for the benefit of global trade but also due to a lack of a local community-based alternatives. Global transportation within the supply chain should continue, but the impact of this transportation on our skies via carbon emissions, our oceans via pollution and marine life disturbance, and our local communities needs to be a part of the calculus.

2030 Goals

By 2030, the UN must complete its open-source platform, using blockchain technology, for global collaboration for the MGI and DeFi investment projects. If a universal approach to an open-source platform is not undertaken by the UN, some other trusted international body—possibly through a social entrepreneurship project—needs to make it happen by 2030. Perhaps tech entrepreneurs, such as Elon Musk and Jeff Bezos, or traditional social entrepreneurship funders like Bill Gates would fund an open-source platform. The other goal for the UN or a social entrepreneurship initiative is to build the backbone of the network infrastructure for a community-based platform to operate as an internet-like system by 2030. Rather than connecting this platform over the existing internet, it is networked as an extranet, connected from community to community, networked to large existing structures, but not necessarily related to open public traffic. Again, this is laying down the infrastructure for the new human systems model of collaboration and community.

By 2030, at least 10 percent of project funds for environment-related projects must come from open-source funding or DeFi to enable Smith's version of capital as flowing efficiently to the opportunities of most need and potential. In addition, the flow of funds must begin to move away from the large public and private institutions that are biased toward large institutional solutions through bodies such as the UN, the World Bank, and the IMF, among others. Through funding and the open-source platform, the goal by 2030 is for 25 percent of materials in the world, natural and synthetic, and raw to finished products to be sequenced and populated in this MGI database to enable public access and protect private ownership. Because there will be inherent private rights and ownership conflicts to be settled as the blockchain technology matures, a starting focus should be on public domain materials for the first 25 percent with fewer conflicts. Some organization such as the World Trade Organization must create an interconnected network of these global MGI databases, built on the open-source platform, as noted above, that is without conflict and suspicion, as is the case today in how the internet operates. This is a very different role for the WTO, and like what I'm proposing for the UN, these intergovernmental organizations must transform themselves to be relevant in the 21st century.

Tied to this MGI database will be the beginning of voluntary BOM listings from manufacturers and suppliers that protect their intellectual property. This also enables the public and interested stakeholders to understand ingredients for their health and the viability of the end-to-end supply chain. By 2032, the goal is 10 percent participation in all products produced.

By 2030, significant legislation will be introduced and enacted to achieve a greater balance between the economy and the environment. First, there needs to be some form of encouragement and/or agreement, led by a body such as the WTO, for reform in the definition and use of products and services that includes product design, product fixability, closed-loop supply chains, and community-based systems, and products as a service (PaaS). The goal by 2030 is for all products to either have cost-effective fixability to reduce waste or exist within an accurate closed-loop system of a community-based supply chain. Additionally, 10 percent of today's existing products should evolve to a PaaS model. From a legal and legislative standpoint, water reform, energy rights, and public utilities must happen. By 2030, there needs to be a reduction in energy and food sector subsidies to 75 percent of today's level. To even the playing field, these funds must either be deployed to 21st-century industries or retired altogether. Legislative reform must happen for the public utility companies written for a 20th-century power grid; reforms should focus on optimizing centralized and decentralized energy generation, storage, and distribution. These reforms must pave the way for a community-based hub-and-spoke system for energy and water. Governments must stop the double talk and address these actions that prohibit environmental innovation.

Related to energy and food, the first publicly funded, community-based, centralized-decentralized energy systems will be built in Africa, with a goal of 25 percent of the continent's energy being from renewable sources. Likewise, public funding should focus on implementing vertical farms and other next-generation technologies and systems to reduce the continent's imports and exports by 25 percent, meaning that Africa will be 25 percent less reliant on other regions of the world. In the developed world, the first stage related to agriculture is the use of improved technologies and processes to make large-scale farming more efficient as a hub model without the need for public subsidies. At the same time, technologies and greater effectiveness of private enterprise solutions for power, agriculture, dairy, and meat will begin to happen, organized around the hub-and-spoke models of these community-based systems. By 2030, vertical farming and precise fermentation must become accepted as mainstream solutions to address the global food problems. The Food and Drug Administration will approve cultivated meats for use, and a hub-and-spoke model will be created for designing cultures and the fermentation/finishing process and then distributing them to 3D printers at community-based systems. By 2030, there should be a 10 percent implementation of these community-based systems for energy, agriculture, dairy, and meat in the United States.

By 2030, the world must have a new definition of water beyond the outdated traditional water cycle model. The existing water from precipitation must be better managed by local communities through better processes and systems, with 100 percent wastewater reclamation and 50 percent rainwater harvesting. By 2030, the initial steps will be taken to build networked, community-based approaches that foster rainwater, wastewater, reclaimed water, and nonpotable sources, such as brackish and salinated water. This means that by the year 2030, there will be a strategy and initial infrastructure for near-infinite water, leading to an end to drought. For chemicals and other synthetic materials, the goal is for 10 percent of these materials to be sequenced, redesigned, and managed. The same is the case for e-materials. I want the percentages for synthetic materials and e-materials to start higher than 10 percent, but realistically, this is a massive and complex problem that will take time and cooperation to address. Finally, the goal by 2030 is for 75 percent reuse of metals and 25 percent reuse of minerals, given the viability of their recycling and reuse.

Hopefully, by the year 2025, there will be a recognition that humans will not save the ecosystem and biodiversity by growing more within it and understand how effective and efficient societies can become if they concentrate in suburban/urban areas within networked community-based systems and relieve the remaining land and sea for environmental healing. The first goal should focus on Africa, the continent where poverty and human systems dysfunction most desperately need to be addressed. The challenges in Africa have some roots within their human systems, as is always the case. Still, the most significant problem is a lack of local self-reliance within a system of global supply chains and the geopolitical world order. A substantial investment in Africa from other world regions, namely the G20, is a critical step in stabilizing the economy and the environment; promises have been made but not kept. The next step is beginning to implement the bifurcation of the built environment as separate from natural protected areas; current estimates put those different natural areas at approximately 15 percent of land and 10 percent of the sea, and the goal by 2030 is for both the land and sea to be increased to 20 percent of the total land mass. This is only an incremental improvement from the current state, but significant human systems hurdles must be addressed before we can get there. If this first transition stage succeeds, it will lead to momentum for 2040 and 2050.

Finally, as these community systems are in place, so should the blockchain and 3D printing infrastructure to create these networked, community-based supply chain systems of manufacturing and logistics. The goal by 2030 is for 10 percent of businesses to be networked within these hub-and-spoke systems that include the implementation of blockchain and 3D printing in these communities. Of course, it will require significant education and training to ensure all citizens can succeed in a digital environment, reducing today's digital divide. As supply chains become glocalized through these community-based systems, it is the perfect time to transform existing transportation from fossil fuels to getting on the grid and using electricity that will eventually be 100 percent renewable. By 2030, 50 percent of trucks, 25 percent of rail locomotives, and 50 percent of passenger cars must be electric or hydrogen based, leading to a 25 percent reduction of these vehicles through automation and ride sharing (PaaS). As for ships and planes, a strategy of either nuclear or hydrogen must emerge as the best option, as battery energy shouldn't be assumed to be sufficient to meet the requirements.

2040 Goals

By 2040, the goal is for 25 percent of all communities in participating nations to be within this community-based network and 25 percent of funding to be within a DeFi, open-source approach that favors individual and community entrepreneurs. The MGI will contain all natural and synthetic materials in the database, and 25 percent of all products will be designed using it. There is also a goal that 25 percent of all products will have a documented BOM, meaning that between the MGI and BOM, there is a significant opportunity for improvement in redesigning materials for the environment, closed-loop systems, recycling, and reuse.

In 2040, 25 percent of products will be PaaS, with the remainder of products either redesigned for better sustainability or within a closed-loop and efficient product fixability. Effective reforms must reduce subsidies by 50 percent from 2025 in the energy and agriculture sectors. As for the fossil fuel energy market, the remaining subsidies are for transitioning these companies into the new energy market and leaving the hydrocarbons in the ground. Likewise, the remaining subsidies for the food industry are effective growth yields distributed within the networked, community-based system. These efforts may not be easy or popular, but they are necessary.

Public utility reform and emerging technologies and processes must lead to a goal of 50 percent of the energy grid on a hub-and-spoke system for the G20 countries. Given transformations happening around the world, such as what's happening with the potential of wind power in the North Sea for Europe and solar power in the Sahara Desert, I am optimistic that these goals can be shattered and met much sooner when we understand how important a strategy of a hub-and-spoke generation and distribution system can be in addressing the challenges. By 2040, the goal is for the G20 countries to get 25 percent of their energy on the grid from renewable sources, which sounds low but will be a challenge while the infrastructure is being developed. However, in the Global South, the goal is 50 percent of a centralized-decentralized grid and 50 percent renewable, with no outages, a game-changing improvement in comparison to the current state.

For the food systems, the goal is for the Global South to have an implementation of 50 percent of the hub-and-spoke communities within these vertical farms and meat and dairy systems through precise fermentation and cultivated meats. In the Global North, the goal is 25 percent. At this point, through a networked system, the goal is that the continent of Africa will be 100 percent self-reliant in terms of its food supply. The continent will still be supported by a hub global supply chain system, but it will be much less reliant on it than in the past. African nations will have gained self-reliance sooner than 2040, but through full autonomy and with support from the global system; these nations will have a more even relationship on the international stage. By 2040, the new model of a water strategy is in place that includes 100 percent rainwater harvesting, 100 percent wastewater reclamation, a 50 percent improvement in the utilization of nonpotable sources, and a networked model so bordering nations and regions of the world are incentivized to cooperate rather than fight over water resources. This will be the year when the world is truly in a situation where water supply is greater than demand.

The planet's clean-up must also take place, with 50 percent of chemicals sequenced, redesigned, or remanaged through containment. By 2040, the earth will begin to see lower levels of trace chemicals in rainwater, wild-caught food such as fish, newborn babies, and remote natural areas. There will be similar goals for e-materials, with 50 percent sequenced, redesigned, and remanaged. For minerals, which are more complicated to recycle than metals but easier than e-materials, the goal will be for 75 percent to be recycled and reused and the remaining 25 percent to be responsibly managed using a closed-loop system or other approaches. Metals must be 100 percent recyclable by 2040.

From 2030 to 2040, funding will transition from Africa to other hot spots in the Global South, such as the small island nations that are hardest hit by climate change. The funding levels for these nations should be sufficient to implement these community-based systems for self-reliance, as was done in Africa. The Global North needs to fund these systems for a more stable global economy and environment through local self-reliance rather than geopolitical interests. This is also a paradigm shift in thinking about how funding is conducted.

Another paradigm shift is the increase by 2040 in the bifurcation between the natural and built environment; the goal is for 30 percent of land and ocean, which is almost double what it is today. By making suburbs and cities more efficient, the natural world will become a healthy place to visit while being protected to improve the lungs of the planet and stabilization of nonhuman species. For supply chains, 25 percent of will be community based in the G20 by 2040, and 100 percent of trucks, cars, and rail locomotives will be electric or hydrogen-based, getting energy from the grid. Emissions from transportation will be brought down significantly using renewable energy feedstocks through a viable grid system and fewer miles driven by community-based supply chains networked to a hub global system. As for commercial ships and planes, either nuclear or green hydrogen will be the dominant energy source, used at 50 percent by 2040. As you can see through these goals, it is the year 2040 when the human systems will become effective in balancing the economy and the planet!

Making It by 2050

If the 2040 goals are accomplished, the next decade will be less intense, as the heavy lifting in the solutions will have been achieved. By 2050, 75 percent of the world's communities of participating nations will be networked in a community-based system; this accounts for the reality that globalization is difficult, if not impossible, and predicting full adherence is unrealistic. The same holds for the financial system, where the goal is 50 percent of the funding for environmental innovation occurring through open-source DeFi markets and the other 50 percent occurring through large, entrenched institutions. By 2050, 75 percent of materials will be designed through the MGI platform, with 100 percent of natural and synthetic materials listed in the database. Only 50 percent of products will be recorded and tracked through BOMs, but I'm hoping consumers will demand greater transparency, leading to higher numbers. The same is true for PaaS of 50 percent, meaning half of the products you buy today will be available as services, resulting in significant de-materialization by 2050. A free-market economist would aspire to reduce all subsidies by 2050; the goal will be a 75 percent reduction, given the nature of human systems. Yet imagine how reducing the hundreds of billions of dollars spent today on protecting sectors of the economy that aren't sustainable could benefit society by liberating opportunities for entrepreneurs who can be more effective in balancing the economy and the environment! Through reforms and transformation, there will be an opportunity for all nations and regions of the world to participate in a hub-and-spoke energy system based on renewable sources. Imagine a networked energy system across Europe, Africa, and Asia that links the enormous wind potential of northern Europe and the solar potential of the Middle East and Africa, distributing energy rationally and efficiently. There is the possibility that some countries primarily invested in fossil fuels or untrusting of their neighbors will decline to participate, but it's hopeful that this won't happen. There is no reason why a centralized-decentralized renewable energy model focused primarily on wind and solar cannot be fully operational by 2050 for any nations that wish to participate.

The same is true for food and water, the most essential resources for life. Inefficient means and inequitable distribution in these areas have led to significant global poverty. The community-based system for food will have industrial production facilities around the world as well as hubs for creating cultures that can be used locally at 3D printing sites for food production. Improvements in seeds and techniques can be perfected glocally and then distributed locally; the same applies to water filtration, reclamation, and yields in use.

Given advances in material sequencing, transparency, and science in design, chemicals and e-materials should be 100 percent sustainable by not being harmful post-use and being contained adequately within the supply chain. Therefore, there should be a ban on synthetic chemicals outside of the supply chain, given it isn't an economic disadvantage anymore to do so. Likewise, all minerals must also be 100 percent sustainable or contained within a closed-loop system alongside metals, without exception.

By 2050, all areas of the Global South should have received funding to create community-based systems, with no strings attached to the G20. The goal, according to some scientists, of 44 percent of the protected environment as a natural ecosystem should be accomplished, with the human population living within the remaining 56 percent and having the opportunity to visit these natural environments without intruding on them. Of course, the oceans must be protected and understood to a greater extent.

Finally, by 2050, all communities should have the option of a community-based supply chain, with a goal of 50 percent participation, networked to the large industrial, global systems. Emerging technologies such as blockchain, 3D printing, quantum computing, and AI will continue to improve and optimize the balance between the local and global. In addition, all vehicles will be electric and run on 100 percent renewable energy, with hydrogen or nuclear fuel used for larger vehicles such as planes and ships.

Finally, you may note that none of these goals for 2030, 2040, and 2050 address CO_2 levels in the sky, ocean plastic levels, and so on. This is because those are symptomatic of the more considerable challenges that are discussed in the twenty strategies. If policymakers, politicians, environmentalists, and scientists chase symptoms, nothing changes; it's only when the root-cause problems are addressed that the magic happens!

Being Human

In 2050, God willing, I will be an old man, still alive. As I think about this, I'm reminded of the *60 Minutes* interview in January 2023 with Paul Ehrlich, a surprisingly fit and coherent man at age ninety, reflecting on a lifetime of scholarly work and predicting the demise of the future, as he has done so many times in the past. Ehrlich was a bit of a celebrity scientist, appearing on Johnny Carson's *Tonight Show* many times and gaining a pop-culture following. Yet if you read the details of his book *The Population Bomb*, there are some predictions that weren't even close—and worse, some unsavory recommendations and implications about controlling the population, such as covert sterilization through the drinking water. While I empathize with an elderly man who has undoubtedly tried to improve the world, even an unintentional strategy that blames the "excess population" for their poverty has proved ineffective and dangerous. The data clearly show how humans can solve these socioeconomic and environmental challenges in tandem rather than in conflict. I am excited about the prospect of watching this transpire through an understanding of human limitations for the betterment of humanity. Our human limitations show that transformation cannot happen through large public and private institutions but must be achieved with individuals and communities as the cornerstones of human advancement and the technology to enable us to better network together. A 21st-century sustainability strategy must also allow humans to relate to nature while at the same time being separate from it.

In this book, I've written about sustainability while primarily focusing on poverty and the limitations of our human species as the root cause of problems. I hope now you understand the significance of this. CO_2 emissions and ocean plastic are symptoms of a more significant root cause, just as relying too much on science, technology, and supply chains isn't the root cause but rather a symptom of the challenges of our human systems. This thought may leave you with some trepidation. It may be hard to fathom not being wholly focused on CO_2 emissions and government policies. However, consider whether this approach has been successful over the past decades. Some may retort that we need more focus. But, of course, that's equivalent to the definition of insanity in doing the same thing repeatedly and expecting a different result. Some may also struggle with acknowledging and accepting our human limitations, but without doing so, we will continue to play Dr. Frankenstein. For the most part, our strategy needs to be focused on being human: not just those of us who consider the environment an existential crisis but also those who cannot afford to focus on 2050, next year, or even tomorrow.

All life on the planet faces the same threats as humans do, but we are the only ones who understand what's happening and that can do something about it. To quote E. O. Wilson, "We have Paleolithic emotions, medieval institutions, and godlike technology."[11] Once we acknowledge this and strategize accordingly, we will have accepted all that is great about our humanity. Our limitations are the result of being an organic life form, meaning we are connected to all life on the planet. As the sole arbiter of life on earth, it's time to redefine our humanity from a more significant and limited perspective within these limitations. It will take courage to think and act differently, but if we can, exciting results will happen sooner rather than later! We can thrive within our human limitations.

Notes

Chapter 1

1. Scott Michon, "What's the Hottest Earth Has Been 'Lately'?" National Oceanic and Atmospheric Administration, September 17, 2014, https://www.climate.gov/news-features/climate-qa/what%E2%80%99s-hottest-earth-has-been-%E2%80%9Clately%E2%80%9D#.

2. Chi Xu, Timothy Kohler, Timothy M. Lenton, Jens-Christian Svenning, and Marten Scheffer, "Future of the Human Climate Niche," *Proceedings of the National Academy of Sciences* 117, no. 21 (2020): 11350–55, https://doi.org/10.1073/pnas.1910114117.

3. Theo Stein, "Carbon Dioxide Now More Than 50% Higher Than Pre-industrial Levels," National Oceanic and Atmospheric Administration, June 3, 2022, https://www.noaa.gov/news-release/carbon-dioxide-now-more-than-50-higher-than-pre-industrial-levels.

4. Noah Harari, *Sapiens: A Brief History of Humankind* (New York: Harper Perennial, 2015).

5. Sebastian Junger, *Tribe: On Homecoming and Belonging* (New York: Twelve, 2016).

6. Emily Westacott, "Understanding Socratic Ignorance," ThoughtCo, February 7, 2019, https://www.thoughtco.com/socratic-ignorance-2670664.

7. Winnie Byanyima, "HIV or COVID-19, Inequity Is Deadly," *Nature Human Behavior* 6 (2022): 176, https://doi.org/10.1038/s41562-022-01307-9.

8. Cary Funk and Meg Hefferon, "U.S. Public Views on Climate and Energy," Pew Research Center, November 25, 2019, https://www.pewresearch.org/science/2019/11/25/u-s-public-views-on-climate-and-energy/.

9. Hallie Golden, "This French Theme Park Doesn't Sugarcoat Its Environmental Message," Bloomberg, April 25, 2019, https://www.bloomberg.com/news/articles/2019-04-25/teaching-kids-about-climate-change-at-a-theme-park.

10. Jonathan Cole, "The Two Cultures Revisited," *Bridge* 26, nos. 3/4 (September 1996), https://www.nae.edu/19579/19582/21020/7358/7432/TheTwoCulturesRevisited.

11. Molly Bergern and Helen Mountford, "Are We Really Ready to Tackle the Climate Crisis? Yes, Here Are 6 Reasons How," World Economic Forum, December 18, 2020, https://www.weforum.org/agenda/2020/12/paris-agreement-climate-change.

12. Vision of Humanity, "2023 Global Peace Index," https://www.visionofhumanity.org/maps/#/.

13. Akihiko Nishio, "When Poverty Meets Climate Change: A Critical Challenge That Demands Cross-Cutting Solutions," World Bank, November 5, 2021, https://blogs.worldbank.org/climatechange/when-poverty-meets-climate-change-critical-challenge-demands-cross-cutting-solutions.

14. Elizabeth Howton, "Nearly Half the World Lives on Less than $5.50 a Day," World Bank, October 18, 2018, https://www.worldbank.org/en/news/press-release/2018/10/17/nearly-half-the-world-lives-on-less-than-550-a-day.

15. Rafael Bernal, "Oxfam: Almost One-Third of US Workers Make Less Than $15 an Hour," *Hill*, March 22, 2022, https://thehill.com/policy/finance/599218-oxfam-almost-one-third-of-us-workers-make-less-than-15-an-hour/.

16. James Bell, Jacob Poushter, Moira Fagan, and Christine Huang, "In Response to Climate Change, Citizens in Advanced Economies Are Willing to Alter How They Live and Work," Pew Research Center, September 14, 2021, https://www.pewresearch.org/global/2021/09/14/in-response-to-climate-change-citizens-in-advanced-economies-are-willing-to-alter-how-they-live-and-work/.

17. R. K. Prabhu and U. R. Rao, *The Mind of Mahatma Gandhi* (Hassell Street Press, 2023).

18. Richard Heinberg, *The End of Growth: Adapting to Our New Economic Reality* (Gabriola Island, BC: New Society Publishers, 2012).

19. Caroline Crist, "Most Americans Report Overwhelming Stress Levels: Poll," WebMD, March 10, 2022, https://www.webmd.com/anxiety-panic/news/20220310/americans-report-overwhelming-stress-poll.

20. Ana Levy-Lyons, "The Banality of Environmental Destruction," *Tikkun* 30, no. 2 (2015), https://muse.jhu.edu/article/578188.

21. Kyle Smith, "Bill Maher vs. Greta Thunberg: 'Shut the F*** Up,'" *National Review*, November 7, 2021, https://www.nationalreview.com/corner/bill-maher-vs-greta-thunberg-shut-the-fk-up/.

22. PA Media, "Greta Thunberg Hits Out at Leaders Who Use Her Fame to 'Look Good,'" *Guardian*, June 27, 2020, https://www.theguardian.com/environment/2020/jun/27/greta-thunberg-hits-out-at-leaders-who-use-her-fame-to-look-good.

23. António Guterres, "Carbon Neutrality by 2050: The World's Most Urgent Mission," United Nations, December 11, 2020, https://www.un.org/sg/en/content/sg/articles/2020-12-11/carbon-neutrality-2050-the-world%E2%80%99s-most-urgent-mission.

24. European Parliament, "What Is Carbon Neutrality and How Can It Be Achieved by 2050?" June 25, 2021, https://www.europarl.europa.eu/news/en/headlines/society/20190926STO62270/what-is-carbon-neutrality-and-how-can-it-be-achieved-by-2050.

25. United Nations Development Programme, "Sustainable Development Goals," https://www.undp.org/sustainable-development-goals.

26. *Economist*, "169 Commandments," March 26, 2015, https://www.economist.com/leaders/2015/03/26/the-169-commandments.

27. Megan Brenan, "Americans' Confidence in Major U.S. Institutions Dips," Gallup, July 14, 2021, https://news.gallup.com/poll/352316/americans-confidence-major-institutions-dips.aspx.

28. United Nations, "Trust in Public Institutions: Trends and Implications for Economic Security," July 20, 2021, https://www.un.org/development/desa/dspd/2021/07/trust-public-institutions/.

29. Richard Wike, "International Views of the UN Are Mostly Positive," Pew Research Center, September 16, 2022, https://www.pewresearch.org/fact-tank/2022/09/16/international-views-of-the-un-are-mostly-positive/.

30. Gallup, "Big Business," 2022, https://news.gallup.com/poll/5248/big-business.aspx.

31. Felix Salmon, "America's Continued Move toward Socialism," Axios, June 25, 2021, https://www.axios.com/2021/06/25/amcricas-continued-move-toward-socialism.

32. Andrew Ross Sorkin, "A Free Market Manifesto That Changed the World, Reconsidered," *New York Times*, September 14, 2020, https://www.nytimes.com/2020/09/11/business/dealbook/milton-friedman-doctrine-social-responsibility-of-business.html.

33. Thomas Malthus, *An Essay on the Principle of Population* (1798).

CHAPTER 2

1. Corky Siemaszko, "Famed Walden Pond, Which Inspired Henry David Thoreau, Is Being Killed by Urine," NBC News, April 6, 2018, https://www.nbcnews.com/science/environment/famed-walden-pond-which-inspired-henry-david-thoreau-being-killed-n863381.

2. E. O. Wilson, *Biophilia* (Cambridge, MA: Harvard University Press, 1984).

3. Bill McKibben, *The End of Nature* (London: Penguin Classics, 2022).

4. Vaclav Smil, *Energy Transitions: Global and National Perspectives*, 2nd ed. (Santa Barbara, CA: Praeger, 2017), appendix A.

5. Daniel Mahler, Nishant Yonzanruth, Ruth Hill, Christoph Lakner, and Nobuo Yoshida, "Pandemic, Prices, and Poverty," World Bank, 2022, https://blogs.worldbank.org/opendata/pandemic-prices-and-poverty.

6. Marc Jones, "How Hard Could Climate Change Hit the Global Economy, and Where Would Suffer Most?" World Economic Forum, April 29, 2022, https://www.weforum.org/agenda/2022/04/climate-change-global-gdp-risk/.

7. Nicholas LePan, "How Much of Earth's Surface Is Covered by Each Country—in One Graphic," World Economic Forum, January 28, 2021, https://www.weforum.org/agenda/2021/01/earth-surface-ocean-visualization-science-countries-russia-canada-china/.

8. Dina Gilio-Whitaker, "The Problem with the Ecological Indian Stereotype," *Tending the Wild*, KCET, February 7, 2017, https://www.kcet.org/shows/tending-the-wild/the-problem-with-the-ecological-indian-stereotype.

9. Ame Vanorio, "What Native Americans Teach Us about Sustainability," Fox Run Environmental Education Center, June 22, 2020, https://www.foxrunenvironmentaleducationcenter.org/ecopsychology/2020/6/8/what-native-americans-teach-us-about-sustainability.

10. Earthday.org, "The History of Earth Day," https://www.earthday.org/history/.

11. Donald Kettl, "Environmental Policy: The Next Generation," Brookings Institution, October 1, 1998, https://www.brookings.edu/research/environmental-policy-the-next-generation/.

12. Paul Krugman, "Why the Republicans Turned against the Environment," *New York Times*, August 15, 2022, https://www.nytimes.com/2022/08/15/opinion/republicans-environment-climate.html.

13. Nadja Popovich, Livia Albeck-Ripka, and Kendra Pierre-Louis, "The Trump Administration Rolled Back More Than 100 Environmental Rules. Here's the Full List," *New York Times*, January 20, 2021, https://www.nytimes.com/interactive/2020/climate/trump-environment-rollbacks-list.html.

14. United Nations, "Climate: World Getting 'Measurably Closer' to 1.5-Degree Threshold," May 9, 2022, https://news.un.org/en/story/2022/05/1117842.

15. Ivana Kattasova, "Not a Single G20 Country Is in Line with the Paris Agreement on Climate, Analysis Shows," CNN, September 16, 2021, https://www.cnn.com/2021/09/15/world/climate-pledges-insufficient-cat-intl/index.html.

16. United Nations Environment Programme, "Environmental Rule of Law: First Global Report," January 24, 2019, https://www.unep.org/resources/assessment/environmental-rule-law-first-global-report.

17. Daniel C. Esty and Ingrid C. Burke, *A Better Planet: 40 Big Ideas for a Sustainable Future* (New Haven, CT: Yale University Press, 2019).

18. Maria Ivanova, "At 50, the UN Environment Programme Must Lead Again," *Nature*, February 16, 2019, https://www.nature.com/articles/d41586-021-00393-5.

19. David Clarke, "EPA Analyzed," *Issues in Science and Technology* 15, no 2 (1999), https://issues.org/clarke-2/.

20. National Center for Charitable Statistics, "The Non-Profit Sector in Brief, 2019," July 20, 2020, https://nccs.urban.org/publication/nonprofit-sector-brief-2019?emulatemode=2#the-nonprofit-sector-in-brief-2019.

21. Noelle Alejandra Salmi, "There Are Too Many Environmental Organizations," Matador Network, July 22, 2020, https://matadornetwork.com/read/too-many-environmental-organizations/.

22. Ibid.

23. Tim Stobierski, "15 Eye-Opening Corporate Social Responsibility Statistics," *Business Insights*, June 15, 2020, https://online.hbs.edu/blog/post/corporate-social-responsibility-statistics.

24. Chandrain Nair, "Why Are Non-scientists Leading the World's Largest Companies' Sustainability Efforts?" *Fortune*, February 23, 2023, https://fortune.com/2023/02/23/why-non-scientists-lead-world-largest-companies-sustainability-efforts-esg-investing-climate-change-greenwashing-politics-chandran-nair/.

25. Greg Petro, "Consumers Demand Sustainable Products and Shopping Formats," *Forbes*, March 11, 2022, https://www.forbes.com/sites/gregpetro/2022/03/11/consumers-demand-sustainable-products-and-shopping-formats/?sh=6a01bbeb6a06.

26. *Economist*, "Three Letters That Won't Save the Planet," July 2022.

27. Robert Reich, "Why Corporate Social Responsibility Is BS," *Guardian*, September 26, 2021, https://www.theguardian.com/commentisfree/2021/sep/26/why-corporate-social-responsibility-is-bs.

CHAPTER 3

1. David Williams, "Adam Smith and Colonialism," *Journal of International Political Theory* 10, no. 3 (2014), https://doi.org/10.1177/1755088214539412.
2. Quoted in Jesse Norman, *Adam Smith: Father of Economics* (New York: Basic Books, 2018).
3. Jesse Norman, "How Adam Smith Would Fix Capitalism," *Financial Times*, June 21, 2018, https://www.ft.com/content/6795a1a0-7476-11e8-b6ad-3823e4384287.
4. Angus Maddison Project, "Global Gross Domestic Product," 2020, http://www.law.uchicago.edu/files/file/coase%20the%20wealth%20of%20nations.pdf.
5. Ronald Bailey and Marian L. Tupy, *Ten Global Trends Every Smart Person Should Know* (Washington, DC: Cato Institute, 2020).
6. David Henderson, "Adam Smith on U.S. Independence," *EconLog*, July 4, 2012, https://www.econlib.org/archives/2012/07/adam_smith_on_u.html.
7. Quoted in Betsy Reed, "Christianity Needs Bishops Who Speak Up for the Poor and Against Oppression," *Guardian*, August 26, 2014, https://www.theguardian.com/world/2014/aug/26/Christianity-needs-bishops-speak-out-against-oppression.
8. United Nations, "Ending Poverty," https://www.un.org/en/global-issues/ending-poverty.
9. World Bank, "Poverty," https://www.worldbank.org/en/topic/poverty.
10. John Creamer, Emily A. Shrider, Kalee Burns, and Frances Chen, "Poverty in the United States," United States Census Bureau, September 13, 2002, https://www.census.gov/library/publications/2022/demo/p60-277.html.
11. USA Facts, "American Poverty in Three Charts," January 21, 2021, https://usafacts.org/articles/american-poverty-in-three-charts/.
12. PR Newswire, "The Number of Consumers Living Paycheck to Paycheck Has Increased Year-over-Year across All Income Levels," August 1, 2022, https://www.prnewswire.com/news-releases/the-number-of-consumers-living-paycheck-to-paycheck-has-increased-year-over-year-across-all-income-levels-301596552.html.
13. Neil Gross, "OP-ED: Is Environmentalism Just for Rich People?" *New York Times*, December 14, 2018, https://www.nytimes.com/2018/12/14/opinion/sunday/yellow-vest-protests-climate.html.
14. Jedediah Britton-Purdy. "Environmentalism Was Once a Social-Justice Movement," *Atlantic*, December 7, 2016, https://www.theatlantic.com/science/archive/2016/12/how-the-environmental-movement-can-recover-its-soul/509831/.
15. Adam Anton, "Trump Launches Presidential Run with Climate Fallacies," E&E News ClimateWire, November 16, 2022, https://www.eenews.net/articles/trump-launches-presidential-run-with-climate-fallacies/.
16. Simon Evans, "Analysis: Which Countries Are Historically Responsible for Climate Change?" *Carbon Brief*, October 5, 2021, https://www.carbonbrief.org/analysis-which-countries-are-historically-responsible-for-climate-change/.

17. *Economist*, "From Inflation to Insurrection," June 25, 2022.

18. Jeffrey Sachs, *The Ages of Globalization: Geography, Technology, and Institutions* (New York: Columbia University Press, 2020).

19. Thomas Friedman, "The World Is Flat After All," *New York Times*, April 3, 2005, https://www.nytimes.com/2005/04/03/magazine/its-a-flat-world-after-all.html.

20. Klaus Schwab. "Globalization 4.0—What Does It Mean?" World Economic Forum, November 5, 2018, https://www.weforum.org/agenda/2018/11/globalization-4-what-does-it-mean-how-it-will-benefit-everyone/.

21. Andrea Willige, "What's Next for Economic Globalization?" World Economic Forum, May 23, 2002, https://www.weforum.org/agenda/2022/05/future-globalization-multilateralism-nationalism-digital-integration/.

22. BBC, "Yellen: 'Global Race to the Bottom' in Corporate Tax," March 23, 2021, https://www.bbc.com/news/business-56500673.

23. Rainer Zitelmann, "Anyone Who Doesn't Know the Following Facts about Capitalism Should Learn Them," *Forbes*, July 27, 2020, https://www.forbes.com/sites/rainerzitelmann/2020/07/27/anyone-who-doesnt-know-the-following-facts-about-capitalism-should-learn-them/?sh=4ce569a73dc1.

24. Thomas Piketty, *Capital and Ideology* (Cambridge, MA: Harvard University Press, 2002).

25. Naomi Oreskes and Erik Conway, *Merchants of Doubt* (London: Bloomsbury, 2010).

26. Benjamin Franta, "What Big Oil Knew about Climate Change, in Its Own Words," *Conversation*, October 28, 2021, https://theconversation.com/what-big-oil-knew-about-climate-change-in-its-own-words-170642.

Chapter 4

1. Paul Feyerabend, *Science in a Free Society* (London: Verso, 1978).

2. Brian Kennedy, Alex Tyson, and Cary Funk, "Americans' Trust in Scientists, Other Groups Declines," Pew Research Center, February 15, 2022, https://www.pewresearch.org/science/2022/02/15/americans-trust-in-scientists-other-groups-declines/.

3. Morgan McFall-Johnsen, "Oppenheimer's Famous Quote 'I Am Become Death' Isn't Really His. The Ominous Words Come from Hindu Scripture," *Business Insider*, July 20, 2023, https://www.businessinsider.com/Oppenheimer-famous-quote-become-death-destroyer-worlds-from-hindu-scripture-2023-7.

4. Carly Little, "When Oppenheimer, 'Father of the Atomic Bomb,' Was Blacklisted," History, May 16, 2023, https://www.history.com/news/father-of-the-atomic-bomb-was-blacklisted-for-opposing-h-bomb.

5. Carly Cassella, "We Have Breached the Safe Planetary Limit for Synthetic Chemicals, Scientists Warn," ScienceAlert, January 19, 2022, https://www.sciencealert.com/synthetic-chemicals-aren-t-just-pushing-earth-s-boundary-they-ve-crossed-it.

6. Scott Sorokin, "Thriving in a World of 'Knowledge Half-Life.'" *CIO*, April 5, 2019, https://www.cio.com/article/219940/thriving-in-a-world-of-knowledge-half-life.html.

7. Science Museum (London), "The 17th Century Society That Transformed Science," August 7, 2019, https://www.sciencemuseum.org.uk/objects-and-stories/17th-century-society-transformed-science.

8. Nicholas Bloom, Charles Jones, John Van Reenen, and Michael Webb, "Are Ideas Getting Harder to Find?" *American Economic Review* 110, no. 4 (2020): 1104–44, https://doi.org/10.1257/aer.20180338.

9. John Horgan, "Opinion: Science Is Running Out of Things to Discover," *National Geographic*, April 9, 2014, https://www.nationalgeographic.com/travel/article/140409-nobel-prize-physics-aging-scientists-string-theory-inflation.

10. John Horgan, *The End of Science* (New York: Basic Books, 2015).

11. Alvin Weinberg, "Impact of Large-Scale Science on the United States," *Science* 134, no. 3473 (1961): 161–64, https://doi.org/10.1126/science.134.3473.161.

12. Derek J. de Solla Price, *Little Science, Big Science* (New York: Columbia University Press, 1963).

13. Donna Lu, "China Overtakes the US in Scientific Research Output," *Guardian*, August 11, 2002, https://www.theguardian.com/world/2022/aug/11/china-overtakes-the-us-in-scientific-research-output.

14. James Mitchell Crow, "US–China Partnerships Bring Strength in Numbers to Big Science Projects," *Nature*, March 22, 2022, https://www.nature.com/nature-index/news-blog/us-china-partnerships-bring-strength-in-numbers-to-big-science-projects.

15. *Economist*, "In Search of a Bright Light," October 29, 2022.

16. Jonathan Gorodischer, "This Day in Jewish History | 1939: Einstein Makes His Biggest Mistake," *Haaretz*, October 11, 2016, https://www.haaretz.com/jewish/2016-10-11/ty-article/1939-einstein-makes-his-biggest-mistake/0000017f-db72-d3a5-af7f-fbfe922c0000.

17. Price, *Little Science, Big Science.*

18. Michael Dine, "How Einstein Arrived at His Theory of General Relativity," Literary Hub, February 10, 2022, https://lithub.com/how-einstein-arrived-at-his-theory-of-general-relativity/.

19. Seyyed Hossein Nasr, *Religion and the Order of Nature* (New York: Oxford University Press, 1996).

20. Soumya Swaminathan, "The WHO's Chief Scientist on a Year of Loss and Learning," *Nature* 588 (2020): 583–85, https://doi.org/10.1038/d41586-020-03556-y.

21. National Science and Technology Council, *Material Genome Strategic Plan*, November 2021, https://www.mgi.gov/sites/default/files/documents/MGI-2021-Strategic-Plan.pdf.

22. Bloom et al., "Are Ideas Getting Harder to Find?"

23. Alistair Nolan, "Artificial Intelligence and the Future of Science," OECD.AI Policy Observatory, October 25, 2021, https://oecd.ai/en/wonk/ai-future-of-science.

24. Glenn Glow, "Environmental Sustainability and AI," *Forbes*, August 21, 2020, https://www.forbes.com/sites/glenngow/2020/08/21/environmental-sustainability-and-ai/.

25. "An Intellectual Entente," *Harvard*, September 10, 2009, https://www.harvardmagazine.com/breaking-news/james-watson-edward-o-wilson-intellectual-entente.

26. Mary Shelley, *Frankenstein* (1918; reprint, Mineola, NY: Dover Publications, 1994).

Chapter 5

1. Natalie Marchant, "The World's Food Waste Problem Is Bigger Than We Thought—Here's What We Can Do about It," World Economic Forum, March 26, 2021, https://www.weforum.org/agenda/2021/03/global-food-waste-solutions/.

2. Jeffery Kluger, "How Fossil Fuels (Really) May Have Killed the Dinosaurs," *Time*, July 14, 2016, https://time.com/4404545/dinosaur-extinctions-fossil-fuels/.

3. Cecil Bohanon, "Economic Recovery: Lessons from the Post–World War II Period," Mercatus Center Policy Brief, September 12, 2012, https://www.mercatus.org/publications/economic-history/economic-recovery-lessons-post-world-war-ii-period.

4. John Kenneth Galbraith, *The Affluent Society*, 40th Anniversary Edition (Boston: Houghton Mifflin, 1998).

5. Thorstein Veblen, *The Theory of the Leisure Class* (Oxford: Oxford University Press, 2009).

6. Business Wire, "Recent Study Reveals More Than a Third of Global Consumers Are Willing to Pay More for Sustainability as Demand Grows for Environmentally-Friendly Alternatives," October 14, 2021, https://www.businesswire.com/news/home/20211014005090/en/Recent-Study-Reveals-More-Than-a-Third-of-Global-Consumers-Are-Willing-to-Pay-More-for-Sustainability-as-Demand-Grows-for-Environmentally-Friendly-Alternatives.

7. Domenico Montanaro, "Poll: Abortion and Inflation Collide as Top Issues in Midterm Elections," National Public Radio, September 8, 2022, https://www.npr.org/2022/09/08/1121535686/poll-abortion-inflation-midterm-elections.

8. Raghuram Rajan, *The Third Pillar: How Markets and the State Leave the Community Behind* (New York: Penguin, 2020).

9. Jack Buffington, *Reinventing the Supply Chain: A 21st-Century Covenant with America* (Washington, DC: Georgetown University Press, 2023).

10. Rajan, *The Third Pillar.*

Chapter 6

1. Royal Society, "Is the Current Level of Atmospheric CO_2 Concentration Unprecedented in Earth's History?" https://royalsociety.org/topics-policy/projects/climate-change-evidence-causes/question-7/.

2. Gregory Wrightstone, *Inconvenient Facts: The Science That Al Gore Doesn't Want You to Know* (San Antonio: Silver Crown Productions, 2017).

3. Annie Sneed, "Ask the Experts: Does Rising CO_2 Benefit Plants?" *Scientific American*, January 23, 2018, https://www.scientificamerican.com/article/ask-the-experts-does-rising-co2-benefit-plants1/.

4. Tais Dahl, "The Impacts of Land Plant Evolution on Earth's Climate and Oxygenation State—An Interdisciplinary Review," *Climate Geology* 547 (2020), https://doi.org/10.1016/j.chemgeo.2020.119665.

5. Michel Penke, "How Nature Helps Fight Climate Change," *Deutsche Welle*, November 25, 2021, https://www.dw.com/en/carbon-sinks-how-nature-helps-fight-climate-change/a-59835700.

6. Bradley Layton, "A Comparison of Energy Densities of Prevalent Energy Sources in Units of Joules per Cubic Meter," *International Journal of Green Energy* 5, no. 6 (2008): 438–55, https://doi.org/10.1080/15435070802498036.

7. University of Michigan Center for Sustainable Systems, "U.S. Energy System Factsheet," 2022, https://css.umich.edu/publications/factsheets/energy/us-energy-system-factsheet.

8. Ibid.

9. US Energy Information Administration, *Annual Energy Outlook,* 2022.

10. *Economist*, "Freedom Lighters," November 12, 2022.

11. British Petroleum, "Statistical Review of World Energy," 2021, https://www.bp.com/content/dam/bp/business-sites/en/global/corporate/pdfs/energy-economics/statistical-review/bp-stats-review-2021-full-report.pdf.

12. Ibid.

13. Jamal Saghir, "Urbanization in Sub-Saharan Africa," Center for Strategic and International Studies, April 12, 2018, https://www.csis.org/analysis/urbanization-sub-saharan-africa.

14. *Economist*, "Powering Africa," November 5, 2022.

15. Dan Welsby, James Price, Steve Pye, and Paul Ekins, "Unextractable Fossil Fuels in a 1.5°C World," *Nature* 597 (2021): 230–24, https://doi.org/10.1038/s41586-021-03821-8.

16. Gregor Semieniuk, Philip Holden, Jean-Francois Mercure, Pablo Salas, Hector Pollitt, Katharine Jobson, Pim Vercoulen, Unnada Chewpreecha, Neil Edwards, and Jorge Viñuales, "Stranded Fossil-Fuel Assets Translate to Major Losses for Investors in Advanced Economies," *Nature Climate Change* 12 (2022): 532–38, https://doi.org/10.1038/s41558-022-01356-y.

17. Jean-Francois Mercure, Pablo Salas, P. Vercoulen, Gregor Semieniuk, A. Lam, Hector Pollitt, Philip Holden, N. Vakifard, Unnada Chewpreecha, Neil Edwards, and Jorge Viñuales (2021).

18. *Economist*, "When Brown Meets Green," December 24, 2022.

19. Matthew Taylor and Jonathan Watts, "Revealed: The 20 Firms behind a Third of All Carbon Emissions," *Guardian*, October 9, 2019, https://www.theguardian.com/environment/2019/oct/09/revealed-20-firms-third-carbon-emissions.

20. *Economist*, "What Is the Fossil-Fuel Industry Doing at COP27?" November 17, 2022, https://www.economist.com/the-economist-explains/2022/11/17/what-is-the-fossil-fuel-industry-doing-at-cop27.

21. Natacha Rousseau, "The Feasibility and Future of Carbon Capture and Storage Technology," Earth.org, May 24, 2022, https://earth.org/the-feasibility-and-future-of-carbon-capture-and-storage-technology/.

22. Rachel Thrasher, Blake Alexander Simmons, and Kyla Tienhaara, "Thousands of Fossil Fuel Projects Are Protected by Treaties. Canceling Them Could Cost Countries Billions," *Fast Company*, May 8, 2022, https://www.fastcompany.com/90750018/how-treaties-protecting-fossil-fuel-investors-could-jeopardize-global-efforts-to-save-the-climate-and-cost-countries-billions.

23. Collin Eaton, "Exxon Hits Record Profits Again as Oil Industry Sees Banner Quarter," *Wall Street Journal*, October 28, 2022, https://www.wsj.com/articles/chevron-reports-huge-profits-as-oil-and-gas-prices-remain-high-11666952101.

24. *Economist*, "Big Oil's New Map," February 11, 2023.

25. Worldometer, "Oil Left in the World," https://www.worldometers.info/oil/.

26. *Economist*, "Docks, Stocks and Many Floating Barrels," September 24, 2022, 17.

27. Lu Soete, "Fossil Fuel Subsidies: How Can We End Our Addiction?" Frontiers Policy Labs, March 3, 2022, https://policylabs.frontiersin.org/content/fossil-fuel-subsidies-how-can-we-end-our-addiction.

28. Johannes Urpelainen and Elisha George, "Reforming Global Fossil Fuel Subsidies: How the United States Can Restart International Cooperation," Brookings Institution, July 14, 2021, https://www.brookings.edu/research/reforming-global-fossil-fuel-subsidies-how-the-united-states-can-restart-international-cooperation/.

29. International Monetary Fund, "Climate Change: Fossil Fuel Subsidies," https://www.imf.org/en/Topics/climate-change/energy-subsidies.

30. Smithsonian Institution, "Emergence of Electrical Utilities in America," National Museum of American History, https://americanhistory.si.edu/powering/past/h1main.htm.

31. Anthropocene Institute, "Electrify Everything," https://anthropoceneinstitute.com/innovations/grid/.

32. *Economist*, "Hug Pylons, Not Trees," April 8, 2023.

33. Laura Feinstein and Eric de Place, "Playing Monopoly; or, How Utilities Make Money," Sightline Institute, May 18, 2020, https://www.sightline.org/2020/05/18/playing-monopoly-or-how-utilities-make-money/.

34. *Economist*, "Why Sun and Wind Need Harnessing," September 3, 2022, 41.

35. Francesco Guarascio and Phuong Nguyen, "Vietnam Boosts Coal Use Plan for 2030 as G7 Climate Offer Stalls," Reuters, November 23, 2022, https://www.reuters.com/business/cop/vietnam-boosts-coal-use-plan-2030-g7-climate-offer-stalls-2022-11-23/.

36. *Economist*, "Skunk No More," October 22, 2022, 17.

37. *Economist*, "Transitional Justice," October 15, 2020.

38. Damian Carrington, "Fossil Fuel Industry Gets Subsidies of $11m a Minute, IMF Finds," *Guardian*, October 6, 2021, https://www.theguardian.com/environment/2021/oct/06/fossil-fuel-industry-subsidies-of-11m-dollars-a-minute-imf-finds.

39. Jocelyn Timperley, "Why Fossil Fuel Subsidies Are So Hard to Kill," *Nature* 598 (2021): 403–5, https://doi.org/10.1038/d41586-021-02847-2.

40. US Energy Information Administration, "Use of Energy Explained," updated June 29, 2023, https://www.eia.gov/energyexplained/use-of-energy/.

41. Solar Energy Industries Association, "U.S. Solar Market Insight," September 8, 2023, https://www.seia.org/us-solar-market-insight.

42. Tibi Puiu, "Solar Is Now the Cheapest Energy in History," ZME Science, October 9, 2022, https://www.zmescience.com/science/solar-is-now-the-cheapest-energy-in-history/.

43. Jonah Fisher, "Switching to Renewable Energy Could Save Trillions—Study," BBC, September 13, 2022, https://www.bbc.com/news/science-environment-62892013.

44. *Economist*, "The People's Power," June 25, 2022, 5.

45. Nidhi Sharma, "The 5 Largest Solar Power Plants in the World," Ornate Solar, July 19, 2022, https://ornatesolar.com/blog/the-5-largest-solar-power-plants-in-the-world.

46. Mark Jacobson, Mark Delucchi, Mary Cameron, Stephen Coughlin, Catherine Hay, Indu Manogaran, Yanbo Shu, and Anna-Katarina von Krauland, "Impacts of Green New Deal Energy Plans on Grid Stability, Costs, Jobs, Health, and Climate in 143 Countries," *One Earth* 1, no. 4 (2019): 449–63, https://doi.org/10.1016/j.oneear.2019.12.003.

47. Paul Hannon, "Russia's War in Ukraine to Cost Global Economy $2.8 Trillion, OECD Says," *Wall Street Journal*, September 26, 2022, https://www.wsj.com/articles/russias-war-in-ukraine-to-cost-global-economy-2-8-trillion-oecd-says-11664177401.

48. Richard Matthews, "Fact Check Reveals Nuclear Energy Is Safe and Clean but Not Renewable," Change Oracle, June 13, 2022, https://changeoracle.com/2022/06/13/fact-check-reveals-nuclear-energy-is-safe-and-clean-but-not-renewable/.

49. BBC, "Germany Extends Nuclear Power amid Energy Crisis," October 18, 2022, https://www.bbc.com/news/world-europe-63294697.

50. International Atomic Energy Agency, "Nuclear Share of Electricity Generation in 2021," IAEA Power Reactor Information System, https://pris.iaea.org/pris/worldstatistics/nuclearshareofelectricitygeneration.aspx.

51. *Economist*, "Pint-Sized Power Stations," March 26, 2022.

52. Catherine Clifford, "Oklo Has a Plan to Make Tiny Nuclear Reactors That Run off Nuclear Waste," CNBC, June 28, 2021, https://www.cnbc.com/2021/06/28/oklo-planning-nuclear-micro-reactors-that-run-off-nuclear-waste.html.

53. Dimitris Mavrokefalidis, "US Startup Unveils Battery Made from Nuclear Waste That Could Last up to 28,000 Years," *Energy Live News*, September 2, 2020, https://www.energylivenews.com/2020/09/02/us-startup-unveils-battery-made-from-nuclear-waste-that-could-last-up-to-28000-years/.

54. Michael Webber and Joshua Rhodes, "The Solution to America's Energy Waste Problem Is Electrification," *UT News*, December 19, 2017, https://news.utexas.edu/2017/12/19/the-solution-to-americas-energy-waste-problem/.

55. Ibid.

56. Crystal Ponti, "'Vampire Energy' Is Sucking the Life out of Our Planet," *Wired*, April 22, 2022, https://www.wired.com/story/vampire-energy-climate-environment-earth-day/.

57. Earthday.org, "A Billion Acts of Green: Slaying Vampire Energy," August 15, 2013, https://www.earthday.org/a-billion-acts-of-green-slaying-vampire-energy/.

58. Markus Pflitsch, "Will Quantum Technology Be the Silver Bullet for Climate Change?" *Forbes*, September 2, 2022, https://www.forbes.com/sites/forbestechcouncil/2022/09/02/will-quantum-technology-be-the-silver-bullet-for-climate-change/.

59. Lauren Everett, "The Future of Energy Storage," *Lab Manager*, March 2, 2022, https://www.labmanager.com/ask-the-expert/the-future-of-energy-storage-27649.

60. Peter Kelly-Detwiler. "Vaulting into Global Energy Storage Markets: Energy Vault Is a Player to Reckon With," *Forbes*, December 2, 2021, https://www.forbes.com/sites/peterdetwiler/2021/12/02/vaulting-into-global-energy-storage-markets-energy-vault-is-a-player-to-reckon-with/?sh=431dccd0217b.

Chapter 7

1. Nathalie Louisgrand, "The Forgotten History of the World's First Restaurant," *Fast Company*, August 27, 2021, https://www.fastcompany.com/90669668/the-forgotten-history-of-the-worlds-first-restaurant.

2. Sylvia Neely, *A Concise History of the French Revolution*, Critical Issues in World and International History (Lanham, MD: Rowman and Littlefield, 2007).

3. *Economist*, "No Food on the Table," November 18, 2022.

4. *Economist*, "Bread-Blocking Bandits," November 5, 2022.

5. *Mumbai Mirror*, "Rising Food Prices Due to Overpopulation, Says UK's Prince Phillip," May 12, 2008, https://mumbaimirror.indiatimes.com/news/world/felix-fritzl-making-amazing-recovery/articleshow/15810271.cms.

6. Zac Ntim, "Prince Philip Quote About Reincarnating as a Deadly Virus to Solve 'Overpopulation' Resurfaces," *Insider*, April 9, 2021, https://www.insider.com/prince-philip-quote-reincarnating-deady-virus-resurfaces-twitter-2021-4.

7. World Wildlife Fund, "WWF Statement on the Death of HRH The Duke of Edinburgh," April 9, 2021, https://wwf.panda.org/wwf_news/?1995941/A-champion-for-the-environment.

8. Gordon Raymer, "Was Prince Philip Really a Climate Warrior or Just a Good Old-Fashioned Conservationist?" *Telegraph*, November 3, 2021, https://www.telegraph.co.uk/royal-family/2021/11/03/prince-philip-really-climate-warrior-just-good-old-fashioned/.

9. Keith Rankin, "Coronavirus—A Malthusian Event?" Scoop, March 27, 2020, https://www.scoop.co.nz/stories/HL2003/S00228/coronavirus-a-malthusian-event.htm.

10. Food and Agriculture Organization of the United Nations, Data Dissemination, https://www.fao.org/statistics/databases/en/; US Government Accountability Office, "Global Food Security: Improved Monitoring Framework Needed to Assess and Report on Feed the Future's Performance," August 31, 2021, https://www.gao.gov/products/gao-21-548.

11. Julian Fulton, Michael Norton, Fraser Schilling. "Water-Indexed Benefits and Impacts of California Almonds," *Ecological Indicators* 96, no. 1 (2019): 711–17, https://doi.org/10.1016/j.ecolind.2017.12.063.

12. Steve Carter, "Looking into the Future: When Will the Earth Run Out of Water?" AquaBliss, September 19, 2022, https://aquabliss.com/blogs/healthy-water/looking-into-the-future-will-the-earth-run-out-of-water-one-day.

13. University of California, Berkeley, "Understanding Global Change: Water Cycle," https://ugc.berkeley.edu/background-content/water-cycle/.

14. *National Geographic*, "A Clean Water Crisis," https://www.nationalgeographic.com/environment/article/freshwater-crisis.

15. Our World in Data, "Freshwater Use by Region," 2015; revised July 2018, https://ourworldindata.org/water-use-stress.

16. Ibid.

17. Boyka Bratanova, Steve Loughnan, Olivier Klein, Almudena Claasen, and Robert Wood, "Poverty, Inequality, and Increased Consumption of Food: Experimental Evidence for a Causal Link," *Appetite* (2016): 162–71, https://doi.org/10.1016/j.appet.2016.01.028.

18. Food and Agriculture Organization of the United Nations, "What Is Happening to Agrobiodiversity?" 2004, https://www.fao.org/3/y5609e/y5609e02.htm.

19. Genevieve Donnellon-May and Zhang Hongzhou, "China's Main Food Security Challenge: Feeding Its Pigs," *Diplomat*, July 6, 2022, https://thediplomat.com/2022/07/chinas-main-food-security-challenge-feeding-its-pigs/.

20. Mihai Andrei, "More Food Is Wasted on Farms Than in Restaurants and Homes Combined," ZME Science, October 4, 2021, https://www.zmescience.com/science/more-food-is-wasted-on-farms-than-in-restaurants-and-homes-combined/.

21. Chuck Abbott, "World Farm Subsidies Hit $2 Billion a Day," *Successful Farming*, June 30, 2020, https://www.agriculture.com/news/business/world-farm-subsidies-hit-2-billion-a-day.

22. Food and Agriculture Organization of the United Nations, "The World Is at a Critical Juncture," 2021, https://www.fao.org/state-of-food-security-nutrition/2021/en/.

23. Michael Kavanaugh, "Household Food Spending Divides the World," *Financial Times*, January 7, 2019, https://www.ft.com/content/cdd62792-0e85-11e9-acdc-4d9976f1533b.

24. Jeremy Erdman, "We Produce Enough Food to Feed 10 Billion People. So Why Does Hunger Still Exist?" Medium, February 1, 2018, https://medium.com/@jeremyerdman/we-produce-enough-food-to-feed-10-billion-people-so-why-does-hunger-still-exist-8086d2657539.

25. George Monbiot, "Why Are We Feeding Crops to Our Cars When People Are Starving?" *Guardian*, June 30, 2022, https://www.theguardian.com/commentisfree/2022/jun/30/crops-cars-starving-biofuels-climate-sustainable.

26. Andrea Miller, "Why the Global Soil Shortage Threatens Food, Medicine and the Climate," CNBC, June 5, 2022, https://www.cnbc.com/2022/06/05/why-the-global-soil-shortage-threatens-food-medicine-and-the-climate.html.

27. United Nations Environment Programme, "A Multi-Billion-Dollar Opportunity: Repurposing Agricultural Support to Transform Food Systems," September 14, 2021, https://www.unep.org/resources/repurposing-agricultural-support-transform-food-systems.

28. Anne Schechinger, "Under Trump, Farm Subsidies Soared and the Rich Got Richer," Environmental Working Group, February 24, 2021, https://www.ewg.org/interactive-maps/2021-farm-subsidies-ballooned-under-trump/.

29. Tariq Khokhar, "Chart: Globally, 70% of Freshwater Is Used for Agriculture," World Bank, March 22, 2017, https://blogs.worldbank.org/opendata/chart-globally-70-freshwater-used-agriculture.

30. Ben Ryder Howe, "Wall Street Eyes Billions in the Colorado's Water," *New York Times*, January 3, 2021, https://www.nytimes.com/2021/01/03/business/colorado-river-water-rights.html.

31. Ibid.

32. *Economist*, "A Time to Plant," November 12, 2022.

33. Christine Whitt, "A Look at America's Family Farms," US Department of Agriculture, July 29, 2021, https://www.usda.gov/media/blog/2020/01/23/look-americas-family-farms.

34. Fortune Business Insights, "With 25.9% CAGR, Vertical Farming Worth USD 20.91 by 2029," GlobeNewswire, July 28, 2022, https://www.globenewswire.com/en/news-release/2022/07/28/2487973/0/en/With-25-9-CAGR-Vertical-Farming-Market-Worth-USD-20-91-Billion-by-2029.html.

35. Joseph Bruneteaux, "Upwards, Not Outwards: The Potential of Vertical Farming," *McGill Business Review*, May 10, 2021, https://mcgillbusinessreview.com/articles/upwards-not-outwards-the-potential-of-vertical-farming.

36. Holly Ober, "Artificial Photosynthesis Can Produce Food without Sunshine." UC Riverside News, June 23, 2022, https://news.ucr.edu/articles/2022/06/23/artificial-photosynthesis-can-produce-food-without-sunshine.

37. Kai Ryssdal and Anais Amin, "How the Dutch Used Technology and Vertical Farming to Became a Major Food Exporter," *Marketplace*, December 1, 2022, https://www.marketplace.org/2022/12/01/dutch-technology-vertical-farming-major-food-exporter/.

38. Hannah Ritchie, "How Much of the World's Land Would We Need on Order to Feed the Global Population with the Average Diet of a Given Country?" Our World in Data, October 3, 2017, https://ourworldindata.org/agricultural-land-by-global-diets.

39. Marcus E. Raichle and Debra A. Gusnard, "Appraising the Brain's Energy Budget," *PNAS* 99, no. 16 (2002): 10237–39, https://doi.org/10.1073/pnas.172399499.

40. Niall McCarthy, "This Study Shows Which Countries Eat the Most Meat," World Economic Forum, May 11, 2020, https://www.weforum.org/agenda/2020/05/the-countries-that-eat-the-most-meat/.

41. Prodigy Press Wire, "Global Plant Based Protein Market Expected to Reach USD 17.4 Billion by 2027—Report by MarketsandMarkets," December 7, 2022, https://www.digitaljournal.com/pr/global-plant-based-protein-market-expected-to-reach-usd-17-4-billion-by-2027-report-by-marketsandmarkets.

42. Jesse Newman, "Beyond Meat's Very Real Problems: Slumping Sausages, Mounting Losses," *Wall Street Journal*, November 21, 2022, https://www.wsj.com/articles/beyond-meat-ethan-brown-stock-layoffs-sausages-11668963839.

43. Nicole Axworthy, "Vegan 'Dairy-Identical' Cheese Made from Animal-Free Casein Will Launch in 2023," Veg News, June 21, 2021, https://vegnews.com/2021/6/vegan-dairy-identical-cheese.

44. Kat Smith, "How Vegan Eggs of the Future Are Made," LiveKindly, June 5, 2021, https://www.livekindly.com/how-vegan-eggs-of-the-future-are-made/.

45. Greg Garrison, Jon Biermacher, and Wade Brorsen, "How Much Will Large-Scale Production of Cell-Cultured Meat Cost?" *Journal of Agriculture and Food Research* 10 (December 2022), https://doi.org/10.1016/j.jafr.2022.100358.

46. PR Newswire, "Over a Third of U.S. Consumers Will Adopt Cultured Meat When Launched, Says New Survey from Future Meat Technologies," October 19, 2021, https://www.prnewswire.com/news-releases/over-a-third-of-us-consumers-will-adopt-cultured-meat-when-launched-says-new-survey-from-future-meat-technologies-301403567.html.

47. Carly Cassella, "Over 50% of Earth's 'Rivers' Actually Stand Still or Run Dry Every Year." ScienceAlert, June 20, 2021, https://www.sciencealert.com/more-than-half-of-the-world-s-rivers-aren-t-flowing-all-the-time.

48. Henry Storey, "Water Scarcity Challenges China's Development Model." *Interpreter*, September 29, 2022, https://www.lowyinstitute.org/the-interpreter/water-scarcity-challenges-china-s-development-model.

49. Jenessa Duncombe, "Geologists Just Revamped the Water Cycle Diagram—with Humans as Headliners," ZME Science, October 14, 2022, https://www.zmescience.com/science/224595/.

50. Echo Xie. "Another Record for China's Seawater Rice with Doubled Yield in 3 Years," *South China Morning Post*, October 14, 2022, https://www.scmp.com/news/china/science/article/3195872/another-record-chinas-seawater-rice-doubled-yield-3-years.

Chapter 8

1. Circle Economy, *The Circularity Gap Report*, 2022, https://drive.google.com/file/d/1NMAUtZcoSLwmHt_r5TLWwB28QJDghi6Q/view.

2. Rahul Rao, "Earth Has More Than 10,000 Kinds of Minerals. This Massive New Catalog Describes Them All," *Popular Science*, July 1, 2022, https://www.popsci.com/science/earth-minerals-catalog/.

3. Ali Somarin, "Ubiquitous Industrial Minerals: Nature's Most Popular Raw Materials," ThermoFisher Scientific, June 3, 2014, https://www.thermofisher.com/blog/mining/ubiquitous-industrial-minerals-natures-most-popular-raw-materials/.

4. Isabella Cota, "How Nations Sitting on Lithium Reserves Are Handling the New 'White Gold' Rush," *El Pais*, February 14, 2022, https://english.elpais.com/economy-and-business/2022-02-14/how-nations-sitting-on-lithium-reserves-are-handling-the-new-white-gold-rush.html.

5. Cullen Hendrix, "Shift to Renewable Energy Could Be a Mixed Blessing for Mineral Exporters," Peterson Institute for International Economics, January 2022, https://www.piie.com/sites/default/files/documents/pb22-1.pdf.

6. Cade Ahlijian, "Congo's Cobalt Contraversy," *globalEDGE*, April 20, 2022, https://globaledge.msu.edu/blog/post/57136/congos-cobalt-controversy.

7. Clare Church and Alec Crawford, "Green Conflict Minerals," International Institute for Sustainable Development, August 2018, https://www.iisd.org/story/green-conflict-minerals/.

8. Greenpeace, "The Myth of Single-Use Plastic Recycling," 2022, https://www.greenpeace.org/usa/the-myth-of-single-use-plastic-and-recycling/.

9. World Bank, "What a Waste 2.0: A Global Snapshot of Waste Management by 2050," https://datatopics.worldbank.org/what-a-waste/trends_in_solid_waste_management.html.

10. National Institutes of Health, "First Complete Sequence of a Human Genome," April 12, 2022, https://www.nih.gov/news-events/nih-research-matters/first-complete-sequence-human-genome.

11. Thomas Kalil and Lloyd Whitman, "The Materials Genome Initiative: The First Five Years," White House, August 2, 2016, https://obamawhitehouse.archives.gov/blog/2016/08/01/materials-genome-initiative-first-five-years.

12. *Materials Today*, "Mapping the Material Genome with Megalibraries and AI," January 7, 2022, https://www.materialstoday.com/nanomaterials/news/mapping-material-genome-megalibraries-ai/.

13. Peter Bentley, "Can We Recycling Concrete?" *BBC Science Focus*, April 20, 2022, https://www.sciencefocus.com/science/can-we-recycle-concrete/.

14. International Aluminum Institute, "Aluminum Recycling Forecast," October 2020, https://international-aluminium.org/resource/aluminium-recycling-fact-sheet/.

15. Organisation for Economic Cooperation and Development, "Plastic Pollution Is Growing Relentlessly as Waste Management and Recycling Fall Short, Says OECD," February 22, 2022, https://www.oecd.org/environment/plastic-pollution-is-growing-relentlessly-as-waste-management-and-recycling-fall-short.htm.

16. Aluminum Association, "Aluminum Cans," https://www.aluminum.org/product-markets/aluminum-cans.

17. Annie White, "Are New Cars and Trucks Getting Heavier?" Capital One, February 14, 2022, https://www.capitalone.com/cars/learn/finding-the-right-car/are-new-cars-and-trucks-getting-heavier/1260.

18. Circle Economy, *The Circularity Gap Report*.

19. Fionn Hargreaves, "50 Items You Would Have to Lug around to Replace Your Smartphone," Daily Mail, October 11, 2017, https://www.dailymail.co.uk/sciencetech/article-4971810/50-items-d-carry-replace-smartphone.html.

20. CSIRO Australia, "Heat Resistant Coral to Fight Bleaching," ScienceDaily, May 21, 2020, https://www.sciencedaily.com/releases/2020/05/200521084728.htm.

Chapter 9

1. Nsikan Akpan, "Which Came First: Society or a Fear Of God?" *PBS NewsHour*, March 20, 2019, https://www.pbs.org/newshour/science/which-came-first-society-or-a-fear-of-god.

2. Ibid.

3. Richard Dawkins, *The Selfish Gene* (London: Oxford University Press, 1976).

4. Harriet Sherwood, "Religion: Why Faith Is Becoming More and More Popular," *Guardian*, August 27, 2018, https://www.theguardian.com/news/2018/aug/27/religion-why-is-faith-growing-and-what-happens-next.

5. Kzai Stastna, "Do Countries Lose Religion as They Gain Wealth?" CBC News, March 12, 2013, https://www.cbc.ca/news/world/do-countries-lose-religion-as-they-gain-wealth-1.1310451.

6. Philip Barker, "Religion and Nationalism in a Modern World," Berkeley Center for Religion, Peace and World Affairs, March 30, 2022, https://berkleycenter.georgetown.edu/responses/religion-and-nationalism-in-a-modern-world.

7. Molly Hanson, "Could Neo-Paganism Be the New 'Religion' of America?" Big Think, September 30, 2019, https://bigthink.com/the-present/modern-paganism/.

8. David Hume, *A Treatise on Human Nature* (East India Publishing Company, 2022).

9. D. Brewer, *Quotes of Mahatma Gandhi, A Words of Wisdom Collection Book* (Lulu, 2019).

10. Richard Wike, "International Views of the UN Are Mostly Positive," Pew Research Center, September 16, 2022, https://www.pewresearch.org/fact-tank/2022/09/16/international-views-of-the-un-are-mostly-positive/.

11. Amina Fayaz, "Democracy in Decline: How BJP Has Caused India's Fall from Freedom," *Brown Political Review*, December 9, 2022, https://brownpoliticalreview.org/2022/12/democracy-in-decline-how-bjp-has-caused-indias-fall-from-freedom/.

12. Christopher Clary, Sameer Lalwani, Niloufer Siddiqui, and Neelanjan Sircar, "Confidence and Nationalism in Modi's India." Stimson Center, August 17, 2022, https://www.stimson.org/2022/confidence-and-nationalism-in-modis-india/.

13. Kevin Roberts, "The European Union as Seen from Washington," Heritage Foundation, December 7, 2022, https://www.heritage.org/europe/commentary/the-european-union-seen-washington.

14. Karl Marx, *The Eighteenth Brumaire of Louis Bonaparte* (Wellred, 2022).

15. Thomas Friedman, "Would Russia or China Help Us If We Were Invaded by Space Aliens?" *New York Times*, November 1, 2021, https://www.nytimes.com/2021/11/01/opinion/climate-glasgow-russia-china.html.

16. *Economist*, "The Committee to Save the Planet," June 11, 2022, 68.

17. Reuters, "Deutsche Bank's DWS Sued by Consumer Group over Alleged Greenwashing," October 24, 2022, https://www.reuters.com/business/finance/deutsche-banks-dws-sued-by-consumer-group-over-alleged-greenwashing-2022-10-24/.

18. Vivienne Walt, "Inside Saudi Arabia's Plan to Go Green While Remaining the World's No. 1 Oil Exporter," *Time*, September 1, 2022, https://time.com/6210210/saudi-arabia-aramco-climate-oil/.

19. McKinsey, "The Net-Zero Transition: What It Would Cost, What It Could Bring," January 2022, https://www.mckinsey.com/capabilities/sustainability/our-insights/the-net-zero-transition-what-it-would-cost-what-it-could-bring.

20. IBIS World, "Global Oil & Gas Exploration & Production—Market Size 2005–2028," March 30, 2022, https://www.ibisworld.com/global/market-size/global-oil-gas-exploration-production/.

21. International Energy Agency, "Renewable Power's Growth Is Being Turbocharged as Countries Seek to Strengthen Energy Security," December 6, 2022, https://www.iea.org/news/renewable-power-s-growth-is-being-turbocharged-as-countries-seek-to-strengthen-energy-security.

22. Lee Clements, *Investing in the Green Economy, 2022: Tracking Growth and Performance in Green Equities*, FTSE Russell, May 2022, https://content.ftserussell.com/sites/default/files/investing_in_the_green_economy_2022_final_8.pdf.

23. International Energy Agency, "Government Energy Spending Tracker," 2022, https://www.iea.org/reports/government-energy-spending-tracker-2/government-energy-spending-tracker#abstract.

24. Adrienne Buller, *The Price of Whales: On the Illusions of Green Capitalism* (Manchester, UK: Manchester University Press, 2022).

25. Chris McGreal, "70 Years and Half a Trillion Dollars Later: What Has the UN Achieved?" *Guardian*, September 7, 2015, https://www.theguardian.com/world/2015/sep/07/what-has-the-un-achieved-united-nations.

26. Joshua Pearce, "Open-Source Powers the United Nations' Sustainability Goals," OpenSource.com, November 30, 2021, https://opensource.com/article/21/11/open-source-un-sustainability.

27. *Economist*, "Africa Will Remain Poor Unless It Uses More Energy," November 3, 2022, https://www.economist.com/middle-east-and-africa/2022/11/03/africa-will-remain-poor-unless-it-uses-more-energy.

28. Peter Neill, "Sovereignty and the Ocean," World Ocean Forum, September 30, 2022, https://thew2o.medium.com/sovereignty-and-the-ocean-87466b8667ea.

29. ***Brandon R. Brown, *Max Planck: Driven by Vision, Broken by War* (Oxford University Press, 2015).

Chapter 10

1. International Energy Agency, "Net Zero by 2050: A Roadmap for the Global Energy Sector," 2021, https://www.iea.org/reports/net-zero-by-2050.

2. Axel Dreher, Jan Egbert-Sturm, and James Vreeland, "Development Aid and International Politics: Does Membership on the UN Security Council Influence World Bank Decisions?" *Journal of Development Economics* 88, no 1 (2009): 1–18, https://doi.org/10.1016/j.jdeveco.2008.02.003.

3. Harry Baker, "Is Drinking Rainwater Safe?" Live Science, August 21, 2022, https://www.livescience.com/is-drinking-rainwater-safe.

4. Tim Worstall, "We Agree, Let's Abolish Subsidies," Adam Smith Institute, February 19, 2021, https://www.adamsmith.org/blog/we-agree-entirely-lets-abolish-subsidies.

5. World Population Review, "Meat Consumption by Country, 2023," https://worldpopulationreview.com/country-rankings/meat-consumption-by-country.

6. Caroline Delbert, "Humans Contain 42 Mystery Chemicals, Which Is Slightly Concerning," *Popular Mechanics*, March 22, 2021, https://www.popularmechanics.com/science/a35903295/mystery-chemicals-in-humans/.

7. Nadia Barboa, Tasha Stoiberb, Olga Naidenko, and David Andrews, "Locally Caught Freshwater Fish across the United States Are Likely a Significant Source of Exposure to PFOS and Other Perfluorinated Compounds," *Environmental Research* 220 (2023), https://www.sciencedirect.com/science/article/abs/pii/S0013935122024926#preview-section-snippets.

8. International Union for Conservation of Nature, "The World Now Protects 15% of Its Land, but Crucial Biodiversity Zones Left Out," press release, September 3, 2016, https://www.iucn.org/news/secretariat/201609/world-now-protects-15-its-land-crucial-biodiversity-zones-left-out.

9. World Conservation Society, "Scientists Show That at Least 44 Percent of Earth's Land Requires Conservation to Safeguard Biodiversity and Ecosystem Services," press release, June 2, 2022, https://newsroom.wcs.org/News-Releases/articleType/ArticleView/articleId/17622/Scientists-Show-that-at-Least-44-Percent-of-Earths-Land-Requires-Conservation-to-Safeguard-Biodiversity-and-Ecosystem-Services.aspx.

10. Peter Thiel, "Competition Is for Losers," *Wall Street Journal*, September 12, 2014, https://www.wsj.com/articles/peter-thiel-competition-is-for-losers-1410535536.

11. Tristan Harris, "Our Brains Are No Match for Our Technology," *New York Times*, December 5, 2019, https://www.nytimes.com/2019/12/05/opinion/digital-technology-brain.html.

Bibliography

Abbott, Chuck. "World Farm Subsidies Hit $2 Billion a Day." *Successful Farming*, June 30, 2020. https://www.agriculture.com/news/business/world-farm-subsidies-hit-2-billion-a-day.

Ahlijian, Cade. "Congo's Cobalt Controversy." *globalEDGE*, April 20, 2022. https://globaledge.msu.edu/blog/post/57136/congos-cobalt-controversy.

Akpan, Nsikan. "Which Came First: Society or a Fear of God?" *PBS NewsHour*, March 20, 2019. https://www.pbs.org/newshour/science/which-came-first-society-or-a-fear-of-god.

Aluminum Association. "Aluminum Cans." https://www.aluminum.org/product-markets/aluminum-cans.

Andrei, Mihai. "More Food Is Wasted on Farms Than in Restaurants and Homes Combined." ZME Science, October 4, 2021. https://www.zmescience.com/science/more-food-is-wasted-on-farms-than-in-restaurants-and-homes-combined/.

Angus Maddison Project. "Global Gross Domestic Product." 2020. http://www.law.uchicago.edu/files/file/coase%20the%20wealth%20of%20nations.pdf

Anthropocene Institute. "Electrify Everything." https://anthropoceneinstitute.com/innovations/grid/.

Anton, Adam. "Trump Launches Presidential Run with Climate Fallacies." E&E News ClimateWire, November 16, 2022. https://www.eenews.net/articles/trump-launches-presidential-run-with-climate-fallacies/.

Axworthy, Nicole. "Vegan 'Dairy-Identical' Cheese Made from Animal-Free Casein Will Launch in 2023." *Veg News*, June 21, 2021. https://vegnews.com/2021/6/vegan-dairy-identical-cheese.

Bailey, Ronald, and Marian L. Tupy. *Ten Global Trends Every Smart Person Should Know*. Washington, DC: Cato Institute, 2020.

Baker, Harry. "Is Drinking Rainwater Safe?" Live Science, August 21, 2022. https://www.livescience.com/is-drinking-rainwater-safe.

Barboa, Nadia, Tasha Stoiberb, Olga Naidenko, and David Andrews. "Locally Caught Freshwater Fish across the United States Are Likely a Significant Source of Exposure to PFOS and Other Perfluorinated Compounds." *Environmental Research* 220 (2023). https://www.sciencedirect.com/science/article/abs/pii/S0013935122024926#preview-section-snippets.

Barker, Philip. "Religion and Nationalism in a Modern World." Berkley Center for Religion, Peace and World Affairs, March 30, 2022. https://berkleycenter.georgetown.edu/responses/religion-and-nationalism-in-a-modern-world,

BBC. "Germany Extends Nuclear Power amid Energy Crisis." October 18, 2022. https://www.bbc.com/news/world-europe-63294697.

———. "Yellen: 'Global Race to the Bottom' in Corporate Tax." March 23, 2021. https://www.bbc.com/news/business-56500673.

Bell, James, Jacob Poushter, Moira Fagan, and Christine Huang. "In Response to Climate Change, Citizens in Advanced Economies Are Willing to Alter How They Live and Work." Pew Research Center, September 14, 2021. https://www.pewresearch.org/global/2021/09/14/in-response-to-climate-change-citizens-in-advanced-economies-are-willing-to-alter-how-they-live-and-work/.

Bentley, Peter. "Can We Recycle Concrete?" *BBC Science Focus*, April 20, 2022. https://www.sciencefocus.com/science/can-we-recycle-concrete/.

Bergern, Molly, and Helen Mountford. "Are We Really Ready to Tackle the Climate Crisis? Yes, Here Are 6 Reasons How." World Economic Forum, December 18, 2020. https://www.weforum.org/agenda/2020/12/paris-agreement-climate-change

Bernal, Rafael, "Oxfam: Almost One-Third of US Workers Make Less Than $15 an Hour." *Hill*, March 22, 2022. https://thehill.com/policy/finance/599218-oxfam-almost-one-third-of-us-workers-make-less-than-15-an-hour/.

Bloom, Nicholas, Charles Jones, John Van Reenen, and Michael Webb. "Are Ideas Getting Harder to Find?" *American Economic Review* 110, no. 4 (2020): 1104–44. https://doi.org/10.1257/aer.20180338.

Bohanon, Cecil. "Economic Recovery: Lessons from the Post–World War II Period." Mercatus Center Policy Brief, September 10, 2012. https://www.mercatus.org/publications/economic-history/economic-recovery-lessons-post-world-war-ii-period.

Boyle, Patrick. "Why Do So Many Americans Distrust Science?" *AAMC News*, May 4, 2002. https://www.aamc.org/news-insights/why-do-so-many-americans-distrust-science.

Bratanova, Boyka, Steve Loughnan, Olivier Klein, Almudena Claasen, and Robert Wood. "Poverty, Inequality, and Increased Consumption of Food: Experimental Evidence for a Causal Link." *Appetite* (2016): 162–71. https://doi.org/10.1016/j.appet.2016.01.028.

Brenan, Megan. "Americans' Confidence in Major U.S. Institutions Dips." Gallup, July 14, 2021. https://news.gallup.com/poll/352316/americans-confidence-major-institutions-dips.aspx.

Brewer, D. *Quotes of Mahatma Gandhi, A Words of Wisdom Collection Book.* N.p.: Lulu, 2019.

British Petroleum. "Statistical Review of World Energy." 2021. https://www.bp.com/content/dam/bp/business-sites/en/global/corporate/pdfs/energy-economics/statistical-review/bp-stats-review-2021-full-report.pdf.

Britton-Purdy, Jedediah. "Environmentalism Was Once a Social-Justice Movement." *Atlantic*, December 7, 2016. https://www.theatlantic.com/science/archive/2016/12/how-the-environmental-movement-can-recover-its-soul/509831/.

Brown, Brandon R. *Planck: Driven by Vision, Broken by War*. Oxford: Oxford University Press, 2015.

Bruneteaux, Joseph. "Upwards, Not Outwards: The Potential of Vertical Farming." *McGill Business Review*, May 10, 2021. https://mcgillbusinessreview.com/articles/upwards-not-outwards-the-potential-of-vertical-farming.

Buffington, Jack. *Reinventing the Supply Chain: A 21st-Century Covenant with America*. Washington, DC: Georgetown University Press, 2023.

Buller, Adrienne. *The Value of a Whale: On the Illusions of Green Capitalism*. Manchester, UK: Manchester University Press, 2022.

Business Wire. "Recent Study Rcveals More Than a Third of Global Consumers Are Willing to Pay More for Sustainability as Demand Grows for Environmentally-Friendly Alternatives." October 14, 2021. https://www.businesswire.com/news/home/20211014005090/en/Recent-Study-Reveals-More-Than-a-Third-of-Global-Consumers-Are-Willing-to-Pay-More-for-Sustainability-as-Demand-Grows-for-Environmentally-Friendly-Alternatives.

Byanyima, Winnie. "HIV or COVID-19, Inequity Is Deadly." *Nature Human Behavior* 6 (2022): 176. https://doi.org/10.1038/s41562-022-01307-9.

Carrington, Damian. "Fossil Fuel Industry Gets Subsidies of $11m a Minute, IMF Finds." *Guardian*, October 6, 2021. https://www.theguardian.com/environment/2021/oct/06/fossil-fuel-industry-subsidies-of-11m-dollars-a-minute-imf-finds.

Carter, Steve. "Looking into the Future: When Will the Earth Run out of Water?" AquaBliss, September 19, 2022. https://aquabliss.com/blogs/healthy-water/looking-into-the-future-will-the-earth-run-out-of-water-one-day.

Cassella, Carly. "Over 50% of Earth's 'Rivers' Actually Stand Still or Run Dry Every Year." ScienceAlert, June 20, 2021. https://www.sciencealert.com/more-than-half-of-the-world-s-rivers-aren-t-flowing-all-the-time.

———. "We Have Breached the Safe Planetary Limit for Synthetic Chemicals, Scientists Warn." ScienceAlert, January 19, 2022. https://www.sciencealert.com/synthetic-chemicals-aren-t-just-pushing-earth-s-boundary-they-ve-crossed-it.

Church, Clare, and Alec Crawford. "Green Conflict Minerals." International Institute for Sustainable Development, August 2018. https://www.iisd.org/story/green-conflict-minerals/.

Circle Economy. *The Circularity Gap Report*. 2022. https://drive.google.com/file/d/1NMAUtZcoSLwmHt_r5TLWwB28QJDghi6Q/view.

Clarke, David. "EPA Analyzed." *Issues in Science and Technology* 15, no. 2 (1999), https://issues.org/clarke-2/.

Clary, Christopher, Sameer Lalwani, Niloufer Siddiqui, and Neelanjan Sircar. "Confidence and Nationalism in Modi's India." Stimson Center, August 17, 2022. https://www.stimson.org/2022/confidence-and-nationalism-in-modis-india/.

Clements, Lee. *Investing in the Green Economy 2022: Tracking Growth and Performance in Green Equities*. FTSE Russell, May 2022. https://content.ftserussell.com/sites/default/files/investing_in_the_green_economy_2022_final_8.pdf.

Clifford, Catherine. "Oklo Has a Plan to Make Tiny Nuclear Reactors That Run off Nuclear Waste." CNBC, June 28, 2021. https://www.cnbc.com/2021/06/28/oklo-planning-nuclear-micro-reactors-that-run-off-nuclear-waste.html.

Cole, Jonathan. "The Two Cultures Revisited." *Bridge* 26, nos. 3/4 (September 1996). https://www.nae.edu/19579/19582/21020/7358/7432/TheTwoCulturesRevisited.

Cota, Isabella. "How Nations Sitting on Lithium Reserves Are Handling the New 'White Gold' Rush." *El Pais*, February 14, 2022. https://english.elpais.com/economy-and-business/2022-02-14/how-nations-sitting-on-lithium-reserves-are-handling-the-new-white-gold-rush.html.

Creamer, John, Emily A. Shrider, Kalee Burns, and Frances Chen. "Poverty in the United States." United States Census Bureau, September 13, 2002. https://www.census.gov/library/publications/2022/demo/p60-277.html.

Crist, Caroline. "Most Americans Report Overwhelming Stress Levels: Poll." WebMD, March 10, 2022. https://www.webmd.com/anxiety-panic/news/20220310/americans-report-overwhelming-stress-poll.

Crow, James Mitchell. "US–China Partnerships Bring Strength in Numbers to Big Science Projects." *Nature*, March 22, 2022. https://www.nature.com/nature-index/news-blog/us-china-partnerships-bring-strength-in-numbers-to-big-science-projects.

CSIRO Australia. "Heat Resistant Coral to Fight Bleaching." ScienceDaily, May 21, 2020. https://www.sciencedaily.com/releases/2020/05/200521084728.htm.

Dahl, Tais. "The Impacts of Land Plant Evolution on Earth's Climate and Oxygenation State—An Interdisciplinary Review." *Climate Geology* 547 (2020): https://doi.org/10.1016/j.chemgeo.2020.119665.

Dawkins, Richard. *The Selfish Gene. Oxford*: Oxford University Press, 1976.

Delbert, Caroline. "Humans Contain 42 Mystery Chemicals, Which Is Slightly Concerning." *Popular Mechanics*, March 22, 2021. https://www.popularmechanics.com/science/a35903295/mystery-chemicals-in-humans/.

Dine, Michael. "How Einstein Arrived at His Theory of General Relativity." Literary Hub, February 10, 2022. https://lithub.com/how-einstein-arrived-at-his-theory-of-general-relativity/.

Donnellon-May, Genevieve, and Zhang Hongzhou. "China's Main Food Security Challenge: Feeding Its Pigs." *Diplomat*, July 6, 2022. https://thediplomat.com/2022/07/chinas-main-food-security-challenge-feeding-its-pigs/.

Dreher, Axel, Jan Egbert-Sturm, and James Vreeland. "Development Aid and International Politics: Does Membership on the UN Security Council Influence World Bank Decisions?" *Journal of Development Economics* 88, no 1 (2009): 1–18. https://doi.org/10.1016/j.jdeveco.2008.02.003.

Duncombe, Jenessa. "Geologists Just Revamped the Water Cycle Diagram—with Humans as Headliners." ZME Science, October 14, 2022. https://www.zmescience.com/science/224595/.

Earthday.org. "A Billion Acts of Green: Slaying Vampire Energy." August 15, 2013. https://www.earthday.org/a-billion-acts-of-green-slaying-vampire-energy/.

.———. "The History of Earth Day." https://www.earthday.org/history/.

Eaton, Collin. "Exxon Hits Record Profits Again as Oil Industry Sees Banner Quarter." *Wall Street Journal*, October 28, 2022. https://www.wsj.com/articles/chevron-reports-huge-profits-as-oil-and-gas-prices-remain-high-11666952101.

Economist. "169 Commandments." March 26, 2015. https://www.economist.com/leaders/2015/03/26/the-169-commandments.

———. "Africa Will Remain Poor Unless It Uses More Energy." November 3, 2022. https://www.economist.com/middle-east-and-africa/2022/11/03/africa-will-remain-poor-unless-it-uses-more-energy.

———. "Big Oil's New Map." February 11, 2023.

———. "Bread-Blocking Bandits." November 5, 2022.

———. "The Committee to Save the Planet." June 11, 2022.

———. "Docks, Stocks and Many Floating Barrels." September 24, 2022.

———. "Freedom Lighters." November 12, 2022.

———. "From Inflation to Insurrection." June 25, 2022.

———. "Hug Pylons, Not Trees." April 8, 2023.

———. "In Search of a Bright Light." October 29, 2022.

———. "No Food on the Table." November 18, 2022.

———. "The People's Power." June 25, 2022.

———. "Pint-Sized Power Stations." March 26, 2022.

———. "Powering Africa." November 5, 2022.

———. "Skunk No More." October 22, 2022.

———. "Three Letters That Won't Save the Planet." July 2022.

———. "A Time to Plant." November 12, 2022.

———. "Transitional Justice." October 15, 2020.

———. "What Is the Fossil-Fuel Industry Doing at COP27?" November 17, 2022. https://www.economist.com/the-economist-explains/2022/11/17/what-is-the-fossil-fuel-industry-doing-at-cop27.

———. "When Brown Meets Green." December 24, 2022.

———. "Why Sun and Wind Need Harnessing." September 3, 2022.

Environmental Working Group. "Biden and Congress Must Reform a Wasteful and Unfair System." February 24, 2021. https://www.ewg.org/interactive-maps/2021-farm-subsidies-ballooned-under-trump/.

Erdman, Jeremy. "We Produce Enough Food to Feed 10 Billion People. So Why Does Hunger Still Exist?" Medium, February 1, 2018. https://medium.com/jeremyerdman/we-produce-enough-food-to-feed-10-billion-people-so-why-does-hunger-still-exist-8086d2657539.

Esty, Daniel C., and Ingrid C. Burke. *A Better Planet: 40 Big Ideas for a Sustainable Future*. New Haven, CT: Yale University Press, 2019.

European Parliament. "What Is Carbon Neutrality and How Can It Be Achieved by 2050?" June 25, 2021. https://www.europarl.europa.eu/news/en/headlines/society/20190926STO62270/what-is-carbon-neutrality-and-how-can-it-be-achieved-by-2050.

Evans, Simon. "Analysis: Which Countries Are Historically Responsible for Climate Change?" *Carbon Brief*, October 5, 2021. https://www.carbonbrief.org/analysis-which-countries-are-historically-responsible-for-climate-change/.

Everett, Lauren. "The Future of Energy Storage." *Lab Manager*, March 2, 2022. https://www.labmanager.com/ask-the-expert/the-future-of-energy-storage-27649.

Fayaz, Amina. "Democracy in Decline: How BJP Has Caused India's Fall from Freedom." *Brown Political Review*, December 9, 2022. https://brownpoliticalreview.org/2022/12/democracy-in-decline-how-bjp-has-caused-indias-fall-from-freedom/.

Feinstein, Laura, and Eric de Place. "Playing Monopoly; or, How Utilities Make Money." Sightline Institute, May 18, 2020. https://www.sightline.org/2020/05/18/playing-monopoly-or-how-utilities-make-money/.

Feyerabend, Paul. *Science in a Free Society*. London: Verso, 1978.

Fisher, Jonah. "Switching to Renewable Energy Could Save Trillions—Study." BBC, September 13, 2022. https://www.bbc.com/news/science-environment-62892013.

Food and Agriculture Organization of the United Nations. Data Dissemination. https://www.fao.org/statistics/databases/en/.

———. "What Is Happening to Agrobiodiversity?" 2004. https://www.fao.org/3/y5609e/y5609e02.htm.

———. "The World Is at a Critical Juncture." 2021. https://www.fao.org/state-of-food-security-nutrition/2021/en/.

Fortune Business Insights. "With 25.9% CAGR, Vertical Farming Worth USD 20.91 by 2029." GlobeNewswire, July 28, 2022. https://www.globenewswire.com/en/news-release/2022/07/28/2487973/0/en/With-25-9-CAGR-Vertical-Farming-Market-Worth-USD-20-91-Billion-by-2029.html.

Franta, Benjamin. "What Big Oil Knew about Climate Change, in Its Own Words." *Conversation*, October 28, 2021. https://theconversation.com/what-big-oil-knew-about-climate-change-in-its-own-words-170642.

Friedman, Thomas. "The World Is Flat After All." *New York Times*, April 3, 2005. https://www.nytimes.com/2005/04/03/magazine/its-a-flat-world-after-all.html.

———. "Would Russia or China Help Us If We Were Invaded by Space Aliens?" *New York Times*, November 1, 2021. https://www.nytimes.com/2021/11/01/opinion/climate-glasgow-russia-china.html.

Fulton, Julian, Michael Norton, and Fraser Schilling. "Water-Indexed Benefits and Impacts of California Almonds." *Ecological Indicators* 96, no. 1 (2019): 711–17. https://doi.org/10.1016/j.ecolind.2017.12.063.

Funk, Cary, and Meg Hefferon. "U.S. Public Views on Climate and Energy." Pew Research Center, November 25, 2019. https://www.pewresearch.org/science/2019/11/25/u-s-public-views-on-climate-and-energy/.

Galbraith, John Kenneth. *The Affluent Society*. 40th Anniversary Edition. Boston: Houghton Mifflin, 1998.

Gallup. "Big Business." 2022. https://news.gallup.com/poll/5248/big-business.aspx.

Garrison, Greg, Jon Biermacher, and Wade Brorsen. "How Much Will Large-Scale Production of Cell-Cultured Meat Cost?" *Journal of Agriculture and Food Research* 10 (December 2022). https://doi.org/10.1016/j.jafr.2022.100358.

Gilio-Whitaker, Dina. "The Problem with the Ecological Indian Stereotype." *Tending the Wild.* KCET, February 7, 2017. https://www.kcet.org/shows/tending-the-wild/the-problem-with-the-ecological-indian-stereotype.

Glow, Glenn. "Environmental Sustainability and AI." *Forbes*, August 21, 2020. https://www.forbes.com/sites/glenngow/2020/08/21/environmental-sustainability-and-ai/.

Golden, Hallie. "This French Theme Park Doesn't Sugarcoat Its Environmental Message." Bloomberg, April 25, 2019. https://www.bloomberg.com/news/articles/2019-04-25/teaching-kids-about-climate-change-at-a-theme-park.

Gorodischer, Johnathan. "This Day in Jewish History | 1939: Einstein Makes His Biggest Mistake." *Haaretz*, October 11, 2016. https://www.haaretz.com/jewish/2016-10-11/ty-article/1939-einstein-makes-his-biggest-mistake/0000017f-db72-d3a5-af7f-fbfe922c0000.

Greenpeace. "The Myth of Single-Use Plastic Recycling." 2022. https://www.greenpeace.org/usa/the-myth-of-single-use-plastic-and-recycling/.

Gross, Neil. "OP-ED: Is Environmentalism Just for Rich People?" *New York Times*, December 14, 2018. https://www.nytimes.com/2018/12/14/opinion/sunday/yellow-vest-protests-climate.html.

Guarascio, Francesco, and Phuong Nguyen. "Vietnam Boosts Coal Use Plan for 2030 as G7 Climate Offer Stalls." Reuters, November 23, 2022. https://www.reuters.com/business/cop/vietnam-boosts-coal-use-plan-2030-g7-climate-offer-stalls-2022-11-23/.

Guterres, António. "Carbon Neutrality by 2050: The World's Most Urgent Mission." United Nations, December 11, 2020. https://www.un.org/sg/en/content/sg/articles/2020-12-11/carbon-neutrality-2050-the-world%E2%80%99s-most-urgent-mission.

Hannon, Paul. "Russia's War in Ukraine to Cost Global Economy $2.8 Trillion, OECD Says." *Wall Street Journal*, September 26, 2022. https://www.wsj.com/articles/russias-war-in-ukraine-to-cost-global-economy-2-8-trillion-oecd-says-11664177401.

Hanson, Molly. "Could Neo-Paganism Be the New 'Religion' of America?" Big Think, September 30, 2019. https://bigthink.com/the-present/modern-paganism/.

Harari, Noah. *Sapiens: A Brief History of Humankind.* New York: Harper Perennial, 2015.

Hargreaves, Fionn. "50 Items You Would Have to Lug around to Replace Your Smartphone." *Daily Mail*, October 11, 2017. https://www.dailymail.co.uk/sciencetech/article-4971810/50-items-d-carry-replace-smartphone.html.

Harris, Tristan. "Our Brains Are No Match for Our Technology." *New York Times*, December 5, 2019. https://www.nytimes.com/2019/12/05/opinion/digital-technology-brain.html.

Harvard. "An Intellectual Entente." September 10, 2009. https://www.harvardmagazine.com/breaking-news/james-watson-edward-o-wilson-intellectual-entente.

Heinberg, Richard. *The End of Growth: Adapting to Our New Economic Reality.* Gabriola Island, BC: New Society Publishers, 2012.

Henderson, David. "Adam Smith on U.S. Independence." *EconLog*, July 4, 2012. https://www.econlib.org/archives/2012/07/adam_smith_on_u.html.

Hendrix, Cullen. "Shift to Renewable Energy Could Be a Mixed Blessing for Mineral Exporters." Peterson Institute for International Economics, January 2022. https://www.piie.com/sites/default/files/documents/pb22-1.pdf.

Horgan, John. *The End of Science*. New York: Basic Books, 2015.

———. "Opinion: Science Is Running Out of Things to Discover." *National Geographic*, April 9, 2014. https://www.nationalgeographic.com/travel/article/140409-nobel-prize-physics-aging-scientists-string-theory-inflation.

Howe, Ben Ryder. "Wall Street Eyes Billions in the Colorado's Water." *New York Times*, January 3, 2021. https://www.nytimes.com/2021/01/03/business/colorado-river-water-rights.html.

Howton, Elizabeth. "Nearly Half the World Lives on Less than $5.50 a Day." World Bank, October 18, 2018. https://www.worldbank.org/en/news/press-release/2018/10/17/nearly-half-the-world-lives-on-less-than-550-a-day.

Hume, David. *A Treatise on Human Nature*. Ottawa, ON: East India Publishing Company, 2022.

IBIS World. "Global Oil and Gas Exploration and Production—Market Size 2005–2028." March 30, 2022. https://www.ibisworld.com/global/market-size/global-oil-gas-exploration-production/.

International Aluminum Institute. "Aluminum Recycling Forecast." October 2020. https://international-aluminium.org/resource/aluminium-recycling-fact-sheet/.

International Atomic Energy Agency. "Nuclear Share of Electricity Generation in 2021." IAEA Power Reactor Information System. https://pris.iaea.org/pris/worldstatistics/nuclearshareofelectricitygeneration.aspx.

International Energy Agency. "Government Energy Spending Tracker." 2022. https://www.iea.org/reports/government-energy-spending-tracker-2/government-energy-spending-tracker#abstract.

———. "Net Zero by 2050: A Roadmap for the Global Energy Sector." 2021. https://www.iea.org/reports/net-zero-by-2050.

———. "Renewable Power's Growth Is Being Turbocharged as Countries Seek to Strengthen Energy Security." December 6, 2022. https://www.iea.org/news/renewable-power-s-growth-is-being-turbocharged-as-countries-seek-to-strengthen-energy-security.

International Monetary Fund. "Climate Change: Fossil Fuel Subsidies." https://www.imf.org/en/Topics/climate-change/energy-subsidies.

International Union for Conservation of Nature. "The World Now Protects 15% of Its Land, but Crucial Biodiversity Zones Left Out." Press release, September 3, 2016. https://www.iucn.org/news/secretariat/201609/world-now-protects-15-its-land-crucial-biodiversity-zones-left-out.

Ivanova, Maria. "At 50, the UN Environment Programme Must Lead Again." *Nature*, February 16, 2019. https://www.nature.com/articles/d41586-021-00393-5.

Jacobson, Mark, Mark Delucchi, Mary Cameron, Stephen Coughlin, Catherine Hay, Indu Manogaran, Yanbo Shu, and Anna-Katarina von Krauland. "Impacts of Green New Deal Energy Plans on Grid Stability, Costs, Jobs, Health, and Climate in 143 Countries." *One Earth* 1, no. 4 (2019): 449–63. https://doi.org/10.1016/j.oneear.2019.12.003.

Jones, Marc. "How Hard Could Climate Change Hit the Global Economy, and Where Would Suffer Most?" World Economic Forum, April 29, 2022. https://www.weforum.org/agenda/2022/04/climate-change-global-gdp-risk/.

Junger, Sebastian. *Tribe: On Homecoming and Belonging.* New York: Twelve, 2016.

Kalil, Thomas, and Lloyd Whitman. "The Materials Genome Initiative: The First Five Years." White House, August 2, 2016. https://obamawhitehouse.archives.gov/blog/2016/08/01/materials-genome-initiative-first-five-years.

Kattasova, Ivana. "Not a Single G20 Country Is in Line with the Paris Agreement on Climate, Analysis Shows." CNN, September 16, 2021. https://www.cnn.com/2021/09/15/world/climate-pledges-insufficient-cat-intl/index.html.

Kavanaugh, Michael. "Household Food Spending Divides the World." *Financial Times*, January 7, 2019. https://www.ft.com/content/cdd62792-0e85-11e9-acdc-4d9976f1533b.

Kelly-Detwiler, Peter. "Vaulting into Global Energy Storage Markets: Energy Vault Is a Player to Reckon With." *Forbes* December 2, 2021. https://www.forbes.com/sites/peterdetwiler/2021/12/02/vaulting-into-global-energy-storage-markets-energy-vault-is-a-player-to-reckon-with/?sh=431dccd0217b.

Kennedy, Brian, Alex Tyson, and Cary Funk. "Americans' Trust in Scientists, Other Groups Declines." Pew Research Center, February 15, 2022. https://www.pewresearch.org/science/2022/02/15/americans-trust-in-scientists-other-groups-declines/.

Kettl, Donald. "Environmental Policy: The Next Generation." Brookings Institution, October 1, 1998. https://www.brookings.edu/research/environmental-policy-the-next-generation/.

Khokhar, Tariq. "Chart: Globally, 70% of Freshwater Is Used for Agriculture." World Bank, March 22, 2017. https://blogs.worldbank.org/opendata/chart-globally-70-freshwater-used-agriculture.

Kluger, Jeffrey. "How Fossil Fuels (Really) May Have Killed the Dinosaurs." *Time*, July 14, 2016. https://time.com/4404545/dinosaur-extinctions-fossil-fuels/.

Krugman, Paul. "Why the Republicans Turned against the Environment." *New York Times*, August 15, 2022. https://www.nytimes.com/2022/08/15/opinion/republicans-environment-climate.html.

Layton, Bradley. "A Comparison of Energy Densities of Prevalent Energy Sources in Units of Joules per Cubic Meter." *International Journal of Green Energy* 5, no. 6 (2008): 438–55. https://doi.org/10.1080/15435070802498036.

LePan, Nicholas. "How Much of Earth's Surface Is Covered by Each Country—in One Graphic." World Economic Forum, January 28, 2021. https://www.weforum.org/agenda/2021/01/earth-surface-ocean-visualization-science-countries-russia-canada-china/.

Levy-Lyons, Ana. "The Banality of Environmental Destruction." *Tikkun* 30, no. 2 (2015). https://muse.jhu.edu/article/578188.

Little, Carly. "When Oppenheimer, 'Father of the Atomic Bomb,' Was Blacklisted." History, May 16, 2023. https://www.history.com/news/father-of-the-atomic-bomb-was-blacklisted-for-opposing-h-bomb.

Louisgrand, Nathalie. "The Forgotten History of the World's First Restaurant." *Fast Company*, August 27, 2021. https://www.fastcompany.com/90669668/the-forgotten-history-of-the-worlds-first-restaurant.

Lu, Donna. "China Overtakes the US in Scientific Research Output." *Guardian*, August 11, 2002. https://www.theguardian.com/world/2022/aug/11/china-overtakes-the-us-in-scientific-research-output.

Mahler, Daniel, Nishant Yonzanruth, Ruth Hill, Christoph Lakner, and Nobuo Yoshida. "Pandemic, Prices, and Poverty." World Bank, 2022. https://blogs.worldbank.org/opendata/pandemic-prices-and-poverty.

Malthus, Thomas. *An Essay on the Principle of Population*. 1798.

Marchant, Natalie. "The World's Food Waste Problem Is Bigger Than We Thought—Here's What We Can Do about It." World Economic Forum, March 26, 2021. https://www.weforum.org/agenda/2021/03/global-food-waste-solutions/.

Marx, Karl. *The Eighteenth Brumaire of Louis Bonaparte*. London: Wellred, 2022.

Materials Today. "Mapping the Material Genome with Megalibraries and AI." January 7, 2022. https://www.materialstoday.com/nanomaterials/news/mapping-material-genome-megalibraries-ai/.

Matthews, Richard. "Fact Check Reveals Nuclear Energy Is Safe and Clean but Not Renewable." Change Oracle, June 13, 2022. https://changeoracle.com/2022/06/13/fact-check-reveals-nuclear-energy-is-safe-and-clean-but-not-renewable/.

Mavrokefalidis, Dimitris. "US Startup Unveils Battery Made from Nuclear Waste That Could Last Up to 28,000 Years." *Energy Live News*, September 2, 2020. https://www.energylivenews.com/2020/09/02/us-startup-unveils-battery-made-from-nuclear-waste-that-could-last-up-to-28000-years/.

McCarthy, Niall. "This Study Shows Which Countries Eat the Most Meat." World Economic Forum, May 11, 2020. https://www.weforum.org/agenda/2020/05/the-countries-that-eat-the-most-meat/

McFall-Johnsen, Morgan. "Oppenheimer's Famous Quote 'I Am Become Death' Isn't Really His. The Ominous Words Come from Hindu Scripture." *Business Insider*, July 20, 2023. https://www.businessinsider.com/Oppenheimer-famous-quote-become-death-destroyer-worlds-from-hindu-scripture-2023-7.

McGreal, Chris. "70 Years and Half a Trillion Dollars Later: What Has the UN Achieved?" *Guardian*, September 7, 2015. https://www.theguardian.com/world/2015/sep/07/what-has-the-un-achieved-united-nations.

McKibben, Bill. *The End of Nature*. London: Penguin Classics, 2022.

McKinsey and Company. "The Net-Zero Transition: What It Would Cost, What It Could Bring." January 2022. https://www.mckinsey.com/capabilities/sustainability/our-insights/the-net-zero-transition-what-it-would-cost-what-it-could-bring.

Mercure, Jean-Francois, Pablo Salas, P. Vercoulen, Gregor Semieniuk, A. Lam, Hector Pollitt, Philip Holden, N. Vakifard, Unnada Chewpreecha, Neil Edwards, and Jorge Viñuales. (2021).

Michon, Scott. "What's the Hottest Earth Has Been 'Lately'?" National Oceanic and Atmospheric Administration, September 17, 2014. https://www.climate.gov/news-features/climate-qa/what%E2%80%99s-hottest-earth-has-been-%E2%80%9Clately%E2%80%9D#.

Miller, Andrea. "Why the Global Soil Shortage Threatens Food, Medicine and the Climate." CNBC, June 5, 2022. https://www.cnbc.com/2022/06/05/why-the-global-soil-shortage-threatens-food-medicine-and-the-climate.html.

Monbiot, George. "Why Are We Feeding Crops to Our Cars When People Are Starving?" *Guardian*, June 30, 2022. https://www.theguardian.com/commentisfree/2022/jun/30/crops-cars-starving-biofuels-climate-sustainable.

Montanaro, Domenico. "Poll: Abortion and Inflation Collide as Top Issues in Midterm Elections." *National Public Radio*, September 8, 2022. https://www.npr.org/2022/09/08/1121535686/poll-abortion-inflation-midterm-elections.

Mumbai Mirror. "Rising Food Prices Due to Overpopulation, Says UK's Prince Philip." May 12, 2008. https://mumbaimirror.indiatimes.com/news/world/felix-fritzl-making-amazing-recovery/articleshow/15810271.cms.

Nair, Chandrain. "Why Are Non-scientists Leading the World's Largest Companies' Sustainability Efforts?" *Fortune*, February 23, 2023. https://fortune.com/2023/02/23/why-non-scientists-lead-world-largest-companies-sustainability-efforts-esg-investing-climate-change-greenwashing-politics-chandran-nair/.

Nasr, Seyyed Hossein. *Religion and the Order of Nature*. New York: Oxford University Press, 1996.

National Center for Charitable Statistics. "The Non-Profit Sector in Brief, 2019." July 20, 2020. https://nccs.urban.org/publication/nonprofit-sector-brief-2019?emulatemode=2#the-nonprofit-sector-in-brief-2019.

National Geographic. "A Clean Water Crisis." https://www.nationalgeographic.com/environment/article/freshwater-crisis.

National Institutes of Health. "First Complete Sequence of a Human Genome." April 12, 2022. https://www.nih.gov/news-events/nih-research-matters/first-complete-sequence-human-genome.

National Science and Technology Council. *Material Genome Strategic Plan*. November 2021. https://www.mgi.gov/sites/default/files/documents/MGI-2021-Strategic-Plan.pdf.

Neely, Sylvia. *A Concise History of the French Revolution*. Critical Issues in World and International History. Lanham, MD: Rowman and Littlefield, 2007.

Neill, Peter. "Sovereignty and the Ocean." World Ocean Forum, September 30, 2022. https://thew2o.medium.com/sovereignty-and-the-ocean-87466b8667ea.

Newman, Jesse. "Beyond Meat's Very Real Problems: Slumping Sausages, Mounting Losses." *Wall Street Journal*, November 21, 2022. https://www.wsj.com/articles/beyond-meat-ethan-brown-stock-layoffs-sausages-11668963839.

Nishio, Akihiko. "When Poverty Meets Climate Change: A Critical Challenge That Demands Cross-Cutting Solutions." World Bank, November 5, 2021. https://blogs.worldbank.org/climatechange/when-poverty-meets-climate-change-critical-challenge-demands-cross-cutting-solutions.

Nolan, Alistair. "Artificial Intelligence and the Future of Science." OECD.AI Policy Observatory, October 25, 2021.https://oecd.ai/en/wonk/ai-future-of-science.

Norman, Jesse. *Adam Smith: Father of Economics*. New York: Basic Books, 2018.

———. "How Adam Smith Would Fix Capitalism." *Financial Times*, June 21, 2018. https://www.ft.com/content/6795a1a0-7476-11e8-b6ad-3823e4384287.

Ntim, Zac. "Prince Philip Quote about Reincarnating as a Deadly Virus to Solve 'Overpopulation' Resurfaces." *Insider*, April 9, 2021. https://www.insider.com/prince-philip-quote-reincarnating-deady-virus-resurfaces-twitter-2021-4.

Ober, Holly. "Artificial Photosynthesis Can Produce Food without Sunshine." *UC Riverside News*, June 23, 2022. https://news.ucr.edu/articles/2022/06/23/artificial-photosynthesis-can-produce-food-without-sunshine.

Organisation for Economic Co-operation and Development. "Plastic Pollution Is Growing Relentlessly as Waste Management and Recycling Fall Short, Says OECD." February 22, 2022. https://www.oecd.org/environment/plastic-pollution-is-growing-relentlessly-as-waste-management-and-recycling-fall-short.htm.

Oreskes, Naomi, and Erik M. Conway. *Merchants of Doubt*. London: Bloomsbury, 2010.

Our World in Data. "Freshwater Use by Region." 2015; revised July 2018. https://ourworldindata.org/water-use-stress.

PA Media. "Greta Thunberg Hits Out at Leaders Who Use Her Fame to 'Look Good.'" *Guardian*, June 27, 2020. https://www.theguardian.com/environment/2020/jun/27/greta-thunberg-hits-out-at-leaders-who-use-her-fame-to-look-good.

Pearce, Joshua. "Open Source Powers the United Nations' Sustainability Goals." OpenSource.com, November 30, 2021. https://opensource.com/article/21/11/open-source-un-sustainability.

Penke, Michel. "How Nature Helps Fight Climate Change." *Deutsche Welle*, November 25, 2021. https://www.dw.com/en/carbon-sinks-how-nature-helps-fight-climate-change/a-59835700.

Petro, Greg. "Consumers Demand Sustainable Products and Shopping Formats." *Forbes*, March 11, 2022. https://www.forbes.com/sites/gregpetro/2022/03/11/consumers-demand-sustainable-products-and-shopping-formats.

Piketty, Thomas. *Capital and Ideology*. Cambridge, MA: Harvard University Press, 2002.

Pflitsch, Markus. "Will Quantum Technology Be the Silver Bullet for Climate Change?" *Forbes*, September 2, 2022. https://www.forbes.com/sites/forbestechcouncil/2022/09/02/will-quantum-technology-be-the-silver-bullet-for-climate-change/.

Ponti, Crystal. "'Vampire Energy' Is Sucking the Life out of Our Planet." *Wired*, April 22, 2022. https://www.wired.com/story/vampire-energy-climate-environment-earth-day/.

Popovich, Nadja, Livia Albeck-Ripka, and Kendra Pierre-Louis. "The Trump Administration Rolled Back More Than 100 Environmental Rules. Here's the Full List." *New York Times*, January 20, 2021. https://www.nytimes.com/interactive/2020/climate/trump-environment-rollbacks-list.html.

Price, Derek J. de Solla. *Little Science, Big Science*. New York: Columbia University Press, 1963.

PR Newswire. "The Number of Consumers Living Paycheck to Paycheck Has Increased Year-over-Year across All Income Levels." August 1, 2022. https://www.prnewswire.com/news-releases/the-number-of-consumers-living-paycheck-to-paycheck-has-increased-year-over-year-across-all-income-levels-301596552.html.

———. "Over a Third of U.S. Consumers Will Adopt Cultured Meat When Launched, Says New Survey from Future Meat Technologies." October 19, 2021. https://www.prnewswire.com/news-releases/over-a-third-of-us-consumers-will-adopt-cultured-meat-when-launched-says-new-survey-from-future-meat-technologies-301403567.html.

Prabhu, R. K., and U. R. Rao. *The Mind of Mahatma Gandhi*. Hassell Street Press, 2023.

Prodigy Press Wire. "Global Plant Based Protein Market Expected to Reach USD 17.4 billion by 2027—Report by MarketsandMarkets." December 7, 2022. https://www.digitaljournal.com/pr/global-plant-based-protein-market-expected-to-reach-usd-17-4-billion-by-2027-report-by-marketsandmarkets.

Puiu, Tibi. "Solar Is Now the Cheapest Energy in History." ZME Science, October 9, 2022. https://www.zmescience.com/science/solar-is-now-the-cheapest-energy-in-history/.

Raichle, Marcus E., and Debra A. Gusnard. "Appraising the Brain's Energy Budget." *PNAS* 99, no. 16 (2002): 10237–39. https://doi.org/10.1073/pnas.172399499.

Rajan, Raghuram. *The Third Pillar: How Markets and the State Leave the Community Behind*. New York: Penguin, 2020.

Rankin, Keith. "Coronavirus—A Malthusian Event?" Scoop, March 27, 2020. https://www.scoop.co.nz/stories/HL2003/S00228/coronavirus-a-malthusian-event.htm.

Rao, Rahul. "Earth Has More Than 10,000 Kinds of Minerals. This Massive New Catalog Describes Them All." *Popular Science*, July 1, 2022. https://www.popsci.com/science/earth-minerals-catalog/.

Raymer, Gordon. "Was Prince Philip Really a Climate Warrior or Just a Good Old-Fashioned Conservationist?" *Telegraph*, November 3, 2021. https://www.telegraph.co.uk/royal-family/2021/11/03/prince-philip-really-climate-warrior-just-good-old-fashioned/.

Reed, Betsy. "Christianity Needs Bishops Who Speak Up for the Poor and Against Oppression." *Guardian*, August 26, 2014. https://www.theguardian.com/world/2014/aug/26/Christianity-needs-bishops-speak-out-against-oppression.

Reich, Robert. "Why Corporate Social Responsibility Is BS." *Guardian*, September 26, 2021. https://www.theguardian.com/commentisfree/2021/sep/26/why-corporate-social-responsibility-is-bs.

Reuters. "Deutsche Bank's DWS Sued by Consumer Group over Alleged Greenwashing." October 24, 2022. https://www.reuters.com/business/finance/deutsche-banks-dws-sued-by-consumer-group-over-alleged-greenwashing-2022-10-24/.

Ritchie, Hannah. "How Much of the World's Land Would We Need in Order to Feed the Global Population with the Average Diet of a Given Country?" Our World in Data, October 3, 2017. https://ourworldindata.org/agricultural-land-by-global-diets.

Roberts, Kevin. "The European Union as Seen from Washington." Heritage Foundation, December 7, 2022. https://www.heritage.org/europe/commentary/the-european-union-seen-washington.

Rousseau, Natacha. "The Feasibility and Future of Carbon Capture and Storage Technology." Earth.org, May 24, 2022. https://earth.org/the-feasibility-and-future-of-carbon-capture-and-storage-technology/.

Royal Society. "Is the Current Level of Atmospheric CO_2 Concentration Unprecedented in Earth's History?" https://royalsociety.org/topics-policy/projects/climate-change-evidence-causes/question-7/.

Ryssdal, Kai, and Anais Amin. "How the Dutch Used Technology and Vertical Farming to Become a Major Food Exporter." *Marketplace*, December 1, 2022. https://www.marketplace.org/2022/12/01/dutch-technology-vertical-farming-major-food-exporter/.

Sachs, Jeffrey. *The Ages of Globalization: Geography, Technology, and Institutions.* New York: Columbia University Press, 2020.

Saghir, Jamal. "Urbanization in Sub-Saharan Africa." Center for Strategic and International Studies, April 12, 2018. https://www.csis.org/analysis/urbanization-sub-saharan-africa.

Salmi, Noelle Alejandra. "There Are Too Many Environmental Organizations." Matador Network, July 22, 2020. https://matadornetwork.com/read/too-many-environmental-organizations/.

Salmon, Felix. "America's Continued Move toward Socialism." Axios, June 25, 2021. https://www.axios.com/2021/06/25/americas-continued-move-toward-socialism.

Schechinger, Anne. "Under Trump, Farm Subsidies Soared and the Rich Got Richer." Environmental Working Group, February 24, 2021. https://www.ewg.org/interactive-maps/2021-farm-subsidies-ballooned-under-trump/.

Schwab, Klaus. "Globalization 4.0—What Does It Mean?" World Economic Forum, November 5, 2018. https://www.weforum.org/agenda/2018/11/globalization-4-what-does-it-mean-how-it-will-benefit-everyone/.

Science Museum (London). "The 17th Century Society That Transformed Science." August 7, 2019. https://www.sciencemuseum.org.uk/objects-and-stories/17th-century-society-transformed-science.

Semieniuk, Gregor, Philip Holden, Jean-Francois Mercure, Pablo Salas, Hector Pollitt, Katharine Jobson, Pim Vercoulen, Unnada Chewpreecha, Neil Edwards, and Jorge Viñuales. "Stranded Fossil-Fuel Assets Translate to Major Losses for Investors in Advanced Economies." *Nature Climate Change* 12 (2022): 532–38. https://doi.org/10.1038/s41558-022-01356-y.

Sharma, Nidhi. "The 5 Largest Solar Power Plants in the World." Ornate Solar, July 19, 2022. https://ornatesolar.com/blog/the-5-largest-solar-power-plants-in-the-world.

Shelley, Mary. *Frankenstein.* 1918; reprint, Mineola, NY: Dover Publications, 1994.

Sherwood, Harriet. "Religion: Why Faith Is Becoming More and More Popular." *Guardian*, August 27, 2018. https://www.theguardian.com/news/2018/aug/27/religion-why-is-faith-growing-and-what-happens-next.

Siemaszko, Corky. "Famed Walden Pond, Which Inspired Henry David Thoreau, Is Being Killed by Urine." NBC News, April 6, 2018. https://www.nbcnews.com/science/environment/famed-walden-pond-which-inspired-henry-david-thoreau-being-killed-n863381.

Smil, Vaclav. *Energy Transitions: Global and National Perspectives.* 2nd edition. Santa Barbara, CA: Praeger, 2017.

Smith, Kat. "How Vegan Eggs of the Future Are Made." LiveKindly, June 5, 2021. https://www.livekindly.com/how-vegan-eggs-of-the-future-are-made/.

Smith, Kyle. "Bill Maher vs. Greta Thunberg: 'Shut the F*** Up.'" *National Review*, November 7, 2021. https://www.nationalreview.com/corner/bill-maher-vs-greta-thunberg-shut-the-fk-up/.

Smithsonian Institution. "Emergence of Electrical Utilities in America." National Museum of American History. https://americanhistory.si.edu/powering/past/h1main.htm.

Sneed, Annie. "Ask the Experts: Does Rising CO_2 Benefit Plants?" *Scientific American*, January 23, 2018. https://www.scientificamerican.com/article/ask-the-experts-does-rising-co2-benefit-plants1/.

Soete, Lu. "Fossil Fuel Subsidies: How Can We End Our Addiction?" Frontiers Policy Labs, March 3, 2022. https://policylabs.frontiersin.org/content/fossil-fuel-subsidies-how-can-we-end-our-addiction.

Solar Energy Industries Association. "U.S. Solar Market Insight." Updated September 8, 2023. https://www.seia.org/us-solar-market-insight.

Somarin, Ali. "Ubiquitous Industrial Minerals: Nature's Most Popular Raw Materials." Thermo Fisher Scientific, June 3, 2014. https://www.thermofisher.com/blog/mining/ubiquitous-industrial-minerals-natures-most-popular-raw-materials/.

Sorkin, Andrew Ross. "A Free Market Manifesto That Changed the World, Reconsidered." *New York Times*, September 14, 2020. https://www.nytimes.com/2020/09/11/business/dealbook/milton-friedman-doctrine-social-responsibility-of-business.html.

Sorokin, Scott. "Thriving in a World of 'Knowledge Half-Life.'" *CIO*, April 5, 2019. https://www.cio.com/article/219940/thriving-in-a-world-of-knowledge-half-life.html.

Stastna, Kzai. "Do Countries Lose Religion as They Gain Wealth?" CBC News, March 12, 2013. https://www.cbc.ca/news/world/do-countries-lose-religion-as-they-gain-wealth-1.1310451.

Stein, Theo. "Carbon Dioxide Now More Than 50% Higher Than Pre-industrial Levels." National Oceanic and Atmospheric Administration, June 3, 2022. https://www.noaa.gov/news-release/carbon-dioxide-now-more-than-50-higher-than-pre-industrial-levels.

Stobierski, Tim. "15 Eye-Opening Corporate Social Responsibility Statistics." *Business Insights*, June 15, 2020. https://online.hbs.edu/blog/post/corporate-social-responsibility-statistics.

Stockholm International Peace Research Institute. "World Military Expenditure Passes $2 Trillion for First Time." April 25, 2022. https://www.sipri.org/media/press-release/2022/world-military-expenditure-passes-2-trillion-first-time.

Storey, Henry. "Water Scarcity Challenges China's Development Model." *Interpreter*, September 29, 2022. https://www.lowyinstitute.org/the-interpreter/water-scarcity-challenges-china-s-development-model.

Swaminathan, Soumya. "The WHO's Chief Scientist on a Year of Loss and Learning." *Nature* 588 (2020): 583–85. https://doi.org/10.1038/d41586-020-03556-y.

Taylor, Matthew, and Jonathan Watts. "Revealed: The 20 Firms behind a Third of All Carbon Emissions." *Guardian*, October 9, 2019. https://www.theguardian.com/environment/2019/oct/09/revealed-20-firms-third-carbon-emissions.

Thiel, Peter. "Competition Is for Losers." *Wall Street Journal*, September 12, 2014. https://www.wsj.com/articles/peter-thiel-competition-is-for-losers-1410535536.

Thrasher, Rachel, Blake Alexander Simmons, and Kyla Tienhaara. "Thousands of Fossil Fuel Projects Are Protected by Treaties. Canceling Them Could Cost Countries Billions." *Fast Company*, May 8, 2022. https://www.fastcompany.com/90750018/how-treaties-protecting-fossil-fuel-investors-could-jeopardize-global-efforts-to-save-the-climate-and-cost-countries-billions.

Timperley, Jocelyn. "Why Fossil Fuel Subsidies Are So Hard to Kill." *Nature* 598 (2021): 403–5. https://doi.org/10.1038/d41586-021-02847-2.

United Nations. "Climate: World Getting 'Measurably Closer' to 1.5-Degree Threshold." May 9, 2022. https://news.un.org/en/story/2022/05/1117842.

———. "Ending Poverty." https://www.un.org/en/global-issues/ending-poverty.

———. "Trust in Public Institutions: Trends and Implications for Economic Security." July 20, 2021. https://www.un.org/development/desa/dspd/2021/07/trust-public-institutions/.

United Nations Development Programme. "Sustainable Development Goals." https://www.undp.org/sustainable-development-goals.

United Nations Environment Programme. "Environmental Rule of Law: First Global Report." January 24, 2019. https://www.unep.org/resources/assessment/environmental-rule-law-first-global-report.

———. "A Multi-Billion-Dollar Opportunity: Repurposing Agricultural Support to Transform Food Systems." September 14, 2021. https://www.unep.org/resources/repurposing-agricultural-support-transform-food-systems.

University of California, Berkeley. "Understanding Global Change: Water Cycle." Understanding Global Change. https://ugc.berkeley.edu/background-content/water-cycle/.

University of Michigan Center for Sustainable Systems. "U.S. Energy System Factsheet." 2022. https://css.umich.edu/publications/factsheets/energy/us-energy-system-factsheet.

Urpelainen, Johannes, and Elisha George. "Reforming Global Fossil Fuel Subsidies: How the United States Can Restart International Cooperation." Brookings Institution, July 14, 2021. https://www.brookings.edu/research/reforming-global-fossil-fuel-subsidies-how-the-united-states-can-restart-international-cooperation/.

USA Facts. "American Poverty in Three Charts." January 21, 2021. https://usafacts.org/articles/american-poverty-in-three-charts/.

US Energy Information Administration. *Annual Energy Outlook*. 2022.

———. "Use of Energy Explained." Updated June 29, 2023. https://www.eia.gov/energyexplained/use-of-energy/.

US Government Accountability Office. "Global Food Security: Improved Monitoring Framework Needed to Assess and Report on Feed the Future's Performance." August 31, 2021. https://www.gao.gov/products/gao-21-548.

Vanorio, Ame. "What Native Americans Teach Us about Sustainability." Fox Run Environmental Education Center, June 22, 2020. https://www.foxrunenvironmentaleducationcenter.org/ecopsychology/2020/6/8/what-native-americans-teach-us-about-sustainability.

Veblen, Thorstein. *The Theory of the Leisure Class*. Oxford: Oxford University Press, 2009.

Vision of Humanity. "2023 Global Peace Index." https://www.visionofhumanity.org/maps/#/.

Walt, Vivienne. "Inside Saudi Arabia's Plan to Go Green while Remaining the World's No. 1 Oil Exporter." *Time*, September 1, 2022. https://time.com/6210210/saudi-arabia-aramco-climate-oil/.

Webber, Michael, and Joshua Rhodes. "The Solution to America's Energy Waste Problem is Electrification." *UT News*, December 19, 2017. https://news.utexas.edu/2017/12/19/the-solution-to-americas-energy-waste-problem/.

Weinberg, Alvin. "Impact of Large-Scale Science on the United States." *Science* 134, no. 3473 (1961): 161–64. https://doi.org/10.1126/science.134.3473.161.

Welsby, Dan, James Price, Steve Pye, and Paul Ekins. "Unextractable Fossil Fuels in a 1.5°C World." *Nature* 597 (2021): 230–34. https://doi.org/10.1038/s41586-021-03821-8.

Westacott, Emily. "Understanding Socratic Ignorance." ThoughtCo, February 7, 2019. https://www.thoughtco.com/socratic-ignorance-2670664.

White, Annie. "Are New Cars and Trucks Getting Heavier?" Capital One, February 14, 2022. https://www.capitalone.com/cars/learn/finding-the-right-car/are-new-cars-and-trucks-getting-heavier/1260.

Whitt, Christine. "A Look at America's Family Farms." US Department of Agriculture, July 29, 2021. https://www.usda.gov/media/blog/2020/01/23/look-americas-family-farms.

Wike, Richard. "International Views of the UN Are Mostly Positive." Pew Research Center, September 16, 2022. https://www.pewresearch.org/fact-tank/2022/09/16/international-views-of-the-un-are-mostly-positive/.

Williams, David. "Adam Smith and Colonialism." *Journal of International Political Theory* 10, no. 3 (2014). https://doi.org/10.1177/1755088214539412.

Willige, Andrea "What's Next for Economic Globalization?" World Economic Forum, May 23, 2002. https://www.weforum.org/agenda/2022/05/future-globalization-multilateralism-nationalism-digital-integration/.

Wilson, E. O. *Biophilia*. Cambridge, MA: Harvard University Press, 1984.

World Bank. "Poverty." https://www.worldbank.org/en/topic/poverty.

———. "What a Waste 2.0: A Global Snapshot of Waste Management by 2050." https://datatopics.worldbank.org/what-a-waste/trends_in_solid_waste_management.html.

World Conservation Society. "Scientists Show That at Least 44 Percent of Earth's Land Requires Conservation to Safeguard Biodiversity and Ecosystem Services." Press release, June 2, 2022. https://newsroom.wcs.org/News-Releases/articleType/ArticleView/articleId/17622/Scientists-Show-that-at-Least-44-Percent-of-Earths-Land-Requires-Conservation-to-Safeguard-Biodiversity-and-Ecosystem-Services.aspx.

World Population Review. "Meat Consumption by Country, 2023." https://worldpopulationreview.com/country-rankings/meat-consumption-by-country.

World Wildlife Fund. "WWF Statement on the Death of HRH The Duke of Edinburgh." Press release, April 9, 2021. https://wwf.panda.org/wwf_news/?1995941/A-champion-for-the-environment.

Worldometer. "Oil Left in the World." https://www.worldometers.info/oil/.

Worstall, Tim. "We Agree Entirely, Let's Abolish Subsidies." Adam Smith Institute, February 9, 2022. https://www.adamsmith.org/blog/we-agree-entirely-lets-abolish-subsidies.

Wrightstone, Gregory. *Inconvenient Facts: The Science That Al Gore Doesn't Want You to Know.* San Antonio: Silver Crown Productions, 2017.

Xie, Echo. "Another Record for China's Seawater Rice with Doubled Yield in 3 Years." *South China Morning Post*, October 14, 2022. https://www.scmp.com/news/china/science/article/3195872/another-record-chinas-seawater-rice-doubled-yield-3-years.

Xu, Chi, Timothy A. Kohler, Timothy M. Lenton, Jens-Christian Svenning, and Marten Scheffer. "Future of the Human Climate Niche." *Proceedings of the National Academy of Sciences* 117, no. 21 (2020): 11350–55. https://doi.org/10.1073/pnas.1910114117.

Zitelmann, Rainer. "Anyone Who Doesn't Know the Following Facts about Capitalism Should Learn Them." *Forbes*, July 27, 2020. https://www.forbes.com/sites/rainerzitelmann/2020/07/27/anyone-who-doesnt-know-the-following-facts-about-capitalism-should-learn-them/.

Index

About the Author

Jack Buffington is the program director/professor for the Supply Chain Management program at the University of Denver and the director of supply chain and sustainability practices for First Key Consulting, a global consulting firm in the brewing industry. Jack held various prior leadership roles in manufacturing and supply chain for MolsonCoors. He earned his PhD in supply chain management from Lulea University of Technology in Lulea, Sweden, and a post-doc at the Royal Institute of Technology in Stockholm, Sweden. Jack has published more than twenty peer-reviewed journal articles and seven nonfiction business books, winning various awards. His most recent book is *Reinventing the Supply Chain: A 21st-Century Covenant with America* (2023).